WHAT EVERY MANAGER NEEDS TO KNOW ABOUT HEALTH AND SAFETY

❖

WHAT EVERY MANAGER NEEDS TO KNOW ABOUT HEALTH AND SAFETY

Ron Akass

Gower

First published by Gower in hardback in 1994 as Essential Health and Safety for Managers

This paperback edition published 1995 by
Gower Publishing Limited
Gower House
Croft Road
Aldershot
Hampshire GU11 3HR
England

Gower
Old Post Road
Brookfield
Vermont 05036
USA

British Library Cataloguing in Publication Data
Akass, Ron
What Every Good Manager Needs to Know
About Health and Safety. – New ed
I. Title
363.11094

ISBN 0–566–07332–3 (Hbk)
0–566–07734–5 (Pbk)

Typeset in Garamond and Avant Garde by Bournemouth Colour Graphics, Parkstone, Poole, Dorset and printed in Great Britain at the University Press, Cambridge.

CONTENTS

LIST OF FIGURES

LIST OF TABLES

❖

LIST OF MANAGEMENT ACTION CHECKLISTS

PREFACE

❖

Since World War II the speed of technological innovation has altered the workplace more profoundly than at any time since the Industrial Revolution. These changes are not confined to any one sector, and are to be found across the entire spectrum of business – in commerce and industry alike. The speed and technical complexity of change have generated more and more regulations aimed at maintaining a safe and healthy working environment.

Even before the advent of the European Single Market, many employers were worried at their inability to comprehend, still less comply with, the stream of health and safety requirements emanating from Westminster, particularly at a time when the uncertain state of the economy precluded growth in staff and resources to respond to it. Indeed, the economic circumstances have caused a great many safety-related jobs to be lost in recent years – or the responsibilities added to those of other employees, often bringing them to saturation point – with a consequent reduction in overall performance.

If there was concern about the situation before 1993, the avalanche of health and safety regulation introduced as a consequence of the Single Market has served to exacerbate it. There is a growing mood of resignation: a feeling that it will be impossible to comply with such a mass of detail, much of it a repetition of existing requirements with subtle nuances – and with the prospect of more to come! Thankfully the UK Government is in the process of examining over 7,000 pieces of regulation – including that concerned with workplace health and safety – to ensure that it is all really necessary and provides 'added value'. In the meantime businesses must do their best to comply with the new requirements. Nevertheless, whatever the outcome of this review, the Single Market does affect every aspect of the way we do business – and occupational health and safety is no exception.

This book is not, nor can it be, a simple solution for the problems described. Neither is it an encyclopedia covering the minutiae of health and safety law and regulation; there are excellent publications available that meet this need. Instead the focus is upon the key statutory health and safety requirements that affect every business, highlighting ways that organizations can manage these responsibilities with confidence and the knowledge that they are doing all that is 'reasonably practicable'.

Each of us is employed for the special skills and attributes that we have to offer, not necessarily for our expertise in workplace health and safety; we have job-related objectives to achieve and deadlines to meet. In turn, if our employer's objectives are not realized, the business could fail, making health and safety concerns academic. The aim of this book is to ensure that health and safety becomes intuitive, an everyday part of the workplace routine – that is, an integral part of the job, and not an appendage to it!

USING THE BOOK AS A MANAGEMENT TOOL

At the end of each chapter there is a 'management action checklist' containing questions which enable the reader to review the principal requirements of the regulations or subject covered.

This is not to suggest that every question in the checklist relates to a specific legal duty, although many do. Rather the approach is to encourage the reader to question what his or her company are doing, and perhaps to stimulate action to improve on what exists at present. In many cases locally developed practices and procedures may only require minor modifications, and in others none at all.

What will become clear to the reader is that the recommendations do not imply huge expense – in fact the reverse. There will, of course, be areas where some expense is unavoidable: for example, where it is necessary to appoint a competent safety adviser to comply with the Management of Health and Safety at Work Regulations, which are described in full in Chapter 15. But in most cases, it is more a matter of modifying the way that things are organized and managed, and introducing some checks and balances.

There are three parts to this book. Part One covers the fundamentals of health and safety, including the Health and Safety at Work Act 1974 (HASAWA) which remains the UK's primary health and safety legislation, notwithstanding the EU-driven regulations. Part Two describes legislation made under HASAWA before 1993 which is applicable to most businesses. In this part there are chapters dealing with key health and safety matters – fire, accident reporting and COSHH – as well as a composite chapter covering a number of regulations, many of which are uncomplicated and require few words of explanation. Finally, Part Three deals with the six sets of regulations enacted under the aegis of HASAWA to give effect to EU directives; these regulations are applicable now, although there are transition periods in which to achieve full compliance for three of the six. These regulations each have a chapter to themselves, and as they will become increasingly relevant, have been covered in some detail.

The action checklists may be used by safety professionals, perhaps to influence their managements to take action; by line managers to verify that they are doing all that they should to safeguard their subordinates and others working in their areas; and not least, by company executives, and in particular those with special responsibility for health and safety in the company, to ask all relevant questions.

Occupational health and safety is a field in which women play an increasing part, bringing to it a natural empathy with the subject and a freshness of approach which is long overdue. References to 'he', 'his' and 'him' throughout the book – particularly when citing, or commenting on, statutes or regulations – should therefore be taken as referring to both genders; the use of these terms merely reflects the author's desire to avoid unnecessarily cumbersome prose.

The comments made on the statutes and regulations represent the author's own opinion and are not intended to reflect the official position. Interpretation of the law is a matter for the courts.

The status of original and EU-motivated law/regulation is correct at the time of writing. For reasons explained in the book, the position is subject to change.

As a consequence of the Government's competitive tendering policy, HMSO no longer sell HSC/HSE publications. Priced publications and most free leaflets can be obtained from HSE Books, PO Box 1999, Sudbury, Suffolk CO10 6FS. Tel: 0787 881165. Fax: 0787 313995. Where possible ISBN numbers in our reference sections have been altered to reflect these changes.

The opportunity has been taken in this paperback edition to incorporate a chapter on the Construction (Design and Management) Regulations 1994 (CD&M), which came into effect on 31 March 1995.

R. C. Akass

PART ONE

THE BASICS

❖

1

HEALTH AND SAFETY AT WORK ACT 1974 (HASAWA)

❖

The Health and Safety at Work Act (HASAWA) is a most important piece of legislation for many reasons. Chief among these is that all employees in the UK are covered by it, whereas before 1974 only about half the population enjoyed statutory protection while at work.

The Act states that everyone at work, irrespective of their status in an enterprise, has duties and responsibilities in respect to health and safety. It reduced and simplified the enforcing inspectorates, giving them wide powers but also requiring them to provide advice and guidance to employers to help them comply with their statutory duties. And it introduced the possibility of custodial sentences and unlimited fines for the most serious breaches (see Chapter 5).

Companies with recognized trade unions must accept union-nominated employee members to serve as safety representatives within the company, and if these representatives request it, the employer must establish a safety committee. Sufficient time must be made available for the safety representatives to carry out their duties, and they must be given paid leave to undergo suitable training (see Chapter 6). The requirement for unionized companies to recognize safety representatives and establish safety committees has prompted some non-union firms to establish safety committees too, recognizing their value in improving health and safety in the business and improving morale.

Although the legislation is styled 'The Health and Safety at Work Act', the general duty placed upon all employers by Section 2(1) of the Act stipulates that: 'It shall be the duty of every employer to ensure, so far as is reasonably practicable, the health, safety *and welfare* at work of all his employees' [emphasis added]. The inclusion of the welfare requirement at once widens the spectrum of an employer's responsibilities to cover matters peripheral to mainstream health and safety, such as toilet facilities, washrooms, temperature, ventilation, facilities for storing outdoor clothing, and so on.

Specific welfare requirements appear in the Offices, Shops and Railways

Premises Act 1963 (OSRPA) and the Factories Act 1961. These two Acts deal with offices and factories respectively; although they both pre-date HASAWA, they remain extant today, but will be largely repealed by December 1995, as welfare matters have now been consolidated for all business premises in the Workplace (Health, Safety and Welfare) Regulations 1992 (HSW) (see Chapter 17), which came into force on 1 January 1993 together with five other EU-driven health and safety regulations. Workplaces which were in operation before 31 December 1992 do not have to achieve full compliance with HSW until 1 January 1996. Therefore the 1963 Act and the welfare sections of the 1961 Factories Act have to remain in parallel operation with HSW until the end of 1995.

As the Single European Market came into being on 1 January 1993, most health and safety legislation will now emanate from Brussels, being ratified in the UK by conversion to regulations enacted under HASAWA. Indeed the six sets of EU-derived health and safety regulations that became effective on 1 January 1993 were enacted under HASAWA, which is and will remain the primary health and safety legislation of the United Kingdom. The specific requirements of HASAWA therefore remain in force in their entirety and must be complied with in addition to the new regulations. In many instances, conscientious attention to HASAWA will suffice to comply with the new requirements.

HASAWA contains specific duties and responsibilities for employers, persons in charge of premises, employees, designers, manufacturers, importers, and directors, managers and others in positions of authority. The health and safety responsibilities of each category are described below, with comment where appropriate.

EMPLOYEES' RESPONSIBILITIES

Every person in a business is an employee, whatever their status in the firm. Those in positions of authority have additional health and safety responsibilities, but in the final analysis they are employees, and must comply with the duties of employees set out in Section 7 of HASAWA, and also in Section 8 which applies to everyone. These duties are as follows:

1. *Section 7(a)* To take reasonable care for their own health and safety at work, and that of others who may be affected by their acts or omissions.
   Comment This duty emphasizes the key principle of HASAWA, which is that safety at work is everyone's business, and not just an employer's duty. If an employee is injured at work, any subsequent claim for compensation will take account of the extent to which the claimant's own acts or omissions, if any, contributed to the accident causing injury.
2. *Section 7(b)* To co-operate with their employer in all the things which he does in order to discharge his health and safety responsibilities.
   Comment Examples of breaches of this duty are failure to respond at once to the emergency evacuation signal, or inhibiting compliance by

others. In fact breaches of this kind would contravene items 1 and 2 above, because such actions would jeopardize the health and safety of others.

3. *Section 8* Not intentionally or recklessly to interfere with or misuse anything provided in the interests of health, safety and welfare.
   Comment There are many examples of breaches of this section, two of the most common being discharging a fire extinguisher when no emergency existed, and wedging open fire doors with fire extinguishers – misuses which also have implications for item 1 above since the safety of others could be compromised.

In all these examples those responsible, while recognizing that they had done wrong – perhaps broken company rules – might be surprised to learn that their actions also constituted criminal offences. Such, then, is the level of ignorance of health and safety legislation generally, despite the duty to provide employees with sufficient instruction, information and training.

EMPLOYERS' RESPONSIBILITIES

The responsibilities of employers in respect to health and safety at work appear in Sections 2 and 3 of HASAWA.

SECTION 2(1) THE GENERAL DUTY

'It shall be the duty of every employer to ensure, so far as is reasonably practicable, the health, safety and welfare at work of all his employees.'

Comment The term 'reasonably practicable' has been defined in *Edwards* v. *National Coal Board* (1949). In this case Asquith, LJ, said:

> 'Reasonably practicable' is a narrower term than 'physically possible' and seems to me to imply that a computation must be made by the owner in which the quantum or risk is placed in one scale and the sacrifice involved in the measures necessary for averting the risk (whether in money, time or trouble) is placed in the other, and that, if it be shown that there is a gross disproportion between them – the risk being insignificant in relation to the sacrifice – the defendants discharge the onus on them.

However, if matters do go wrong, it is for the employer to show that the actions he took were reasonably practicable in all the circumstances.

SECTION 2(2)

The general duty in Section 2(1) is amplified by five specific requirements in section 2(2) as follows:

1. *Section 2(2)(a)* The provision and maintenance of plant and systems of work that are, so far as is reasonably practicable, safe and without risks to health.

Comment It should not be thought that this section applies only to industry; it has considerable relevance to offices. For example, most offices have procedures to deal with copier operation and the action to be taken in the event of a breakdown. This is a 'system of work'. The procedure for safe operation, maintenance and independent inspection of lifts is an example of 'plant' in this context with regard to an office; boilers are another.

A safe system of work should be developed where the following applies:

- There is a real risk of injury.
- Where there is some degree of complexity or unfamiliarity with the work.
- Some practical precaution/s are possible.

Safe systems should be developed in liaison with the workers who normally carry out the work if this is possible.

In the event of an externally reportable accident or dangerous occurrence, visiting inspectors will ask for details of the system of work at the commencement of their investigations. This line of enquiry frequently reveals the absence of any kind of systematic analysis of work processes and safe methods of carrying them out.

2. *Section 2(2)(b)* Arrangements for ensuring, so far as is reasonably practicable, safety and absence of risks to health in connection with the use, handling, storage and transport of articles and substances.

   Comment This duty has now been codified with the introduction of the Control of Substances Hazardous to Health Regulations 1988 (see Chapter 10).

3. *Section 2(2)(c)* The provision of such information, instruction, training and supervision as is necessary to ensure, so far as is reasonably practicable, the health and safety at work of his employees.

   Comment See Chapter 8.

4. *Section 2(2)(d)* So far as is reasonably practicable as regards any place of work under the employer's control, the maintenance of it in a condition that is safe and without risks to health and the provision and maintenance of means of access to and egress from it that are safe and without such risks.

   Comment This section emphasizes the importance of maintaining the integrity of exits and exit routes (see Chapters 11 and 17).

5. *Section 2(2)(e)* The provision and maintenance of a working environment for employees that is, so far as is reasonably practicable, safe, without risk to health, and adequate as regards facilities and arrangements for their welfare at work.

   Comment Developments in recent years have more than justified the inclusion of this duty. The smoking issue, legionnaires' disease, sick building syndrome, air and water quality, are all matters falling within the ambit of this subsection.

SECTION 2(3)

This subsection states a requirement to develop and publish a health and safety policy (see Chapter 3).

SECTION 2 (4–7)

These subsections concern the duty of an employer with recognized trade unions to allow them to appoint employee safety representatives and in certain circumstances to require the formation of a safety committee (see Chapter 6).

SECTION 3 DUTY TO OTHERS NOT BEING EMPLOYEES

This section requires employers to ensure that persons not in their employment are not adversely affected by their operations. It also applies to the self-employed. Wherever possible, employers and the self-employed should inform persons that might be adversely affected by their operations of the dangers.

Comment This is an onerous duty, and compliance with it should not be left to chance. The caring employer will keep under regular review the arrangements for ensuring the health and safety of non-employees on his premises. Non-employees will normally fall into three categories:

- Business callers (e.g. representatives, prospective customers).
- Visits by interested parties or dignitaries.
- Contractors carrying out work on the premises.

There is a further group who could be adversely affected by an employer's operations – persons living in proximity to the premises who could be affected by noise, atmospheric pollution or smell. Where this happens, and complaints are made to the relevant authority, this may result in action under other legislation.

However, of the first three categories of non-employee listed, contractors pose by far the greatest problem (see Chapter 7). For the first two categories, there is a need to take reasonable measures to minimize danger. Examples of problems likely to arise are as follows:

- The factory tour during which one of the party slips away to the toilet, and loses contact with the main party.
- Mother bringing new baby to premises to show to former colleagues.
- Older children brought to work – either due to domestic problems at home, or to accompany the parent as a 'treat' when the parent comes into work in the evening or at week-ends.

All of these situations pose potential health and safety problems for employers, the first in factories, the latter two more likely in offices. They are by no means uncommon.

Where babies and children are concerned, taking a strong line is often difficult, either because the offending staff are managers, or they are giving up their time voluntarily to assist with a deadline or the like. Yet whatever the circumstances, no child or young person should be allowed beyond reception. In the event of an accident – a much greater possibility with babies and children than with one's own employees – the employer will have no sustainable defence under Section 2(3), and however much goodwill exists between the parent and their employer, this will evaporate when the question of damages/compensation looms. Furthermore, the insurance carrier will be entitled to reject any claim that arises. Insurers expect that the insured company will follow normal business practice and will act within the law. This apart, the wording of the insurance policy will *not* include the children of employees!

Many employers have produced a simple information card for visitors which contains all the key health and safety information, especially in respect to emergency evacuation. While this does not completely absolve an employer from responsibility for visitors, it is a simple practical step – and demonstrates a responsible attitude. An additional safeguard would be to require visitors to be accompanied by a member of staff at all times.

This is fine in theory, but is often difficult in practice. Sudden visits to the WC can upset the best laid plans. and 'Murphy's law' comes into play (e.g. the fire bells will always sound at the most inconvenient moment!).

SECTION 4

This section covers the duties of persons in charge of premises to persons who are not their employees but use premises over which they have control. The primary duty is to ensure that the premises, and means of access to and egress from it, and any plant or substance in it, or provided for use there, is safe and without risks to health.

SECTION 6

This section deals with the duties of designers, suppliers and manufacturers. Essentially these duties relate to articles and substances designed and manufactured for use by others.

There must be adequate research and testing of the products to ensure that they are safe to use for the purposes prescribed; manufacturers or suppliers of substances must provide purchasers or intending purchasers with all relevant information about their products, for example: any hazardous properties; whether they are 'classified' substances; what precautions are necessary when using them; the first-aid and medical treatment if the substance gets onto the skin or in the eyes, or is inhaled or ingested. The COSHH regulations (see Chapter 10) have created a huge demand for this kind of information, and manufacturers produce product data sheets in which all the relevant information is consolidated.

SECTION 9

This regulation makes it an offence for an employer to charge employees for the provision of any protective clothing or equipment necessary to do their work safely. Similarly, if it is deemed necessary for employees to undergo medical examinations as a precaution, no charge may be levied for these.

SECTION 36

This section is concerned with cases where a health and safety offence is committed by one person as a consequence of the default of another; in such circumstances that other person may be proceeded against, whether or not proceedings are taken against the first person.

SECTION 37

This is a most important section relating to the individual personal responsibilities of directors, managers, and others in positions of authority. It states:

> Where an offence under any of the relevant statutory provisions committed by a body corporate is proved to have been committed with the consent or connivance of, or to have been attributable to any neglect on the part of any director, manager, secretary or other similar officer of the body corporate, or a person who was purporting to act in such capacity, he as well as the body corporate shall be liable to be proceeded against and punished accordingly.

Simply put, this means that if a company commits a health and safety offence, and the blame for the breach – or indeed the cause of it – is found to be that of a director or manager, the individual concerned may be indicted, whether or not proceedings are taken against the company. Clearly this is not only intended to cover breaches by managers at the 'sharp end' but could be used against, for example, a finance director who would not allow funding for necessary safety work, or a director who, although responsible for health and safety in the company, only paid lip-service to it.

In a recent case a company director *and* his company were prosecuted for a breach of a prohibition order; both were convicted and fined. However, the court also disqualified the director from holding a directorship for two years under the Company Directors Disqualification Act 1986. This was the first time a disqualification had occurred under this act for a health and safety offence. Although the Company Directors Disqualification Act was designed to deal with company law offences, it was stated that there was nothing in the wording of the act to prevent a disqualification for a health and safety offence.

OTHER SECTIONS OF HASAWA

Those parts of HASAWA dealing with offences, powers of inspectors and similar matters are covered by Chapter 5.

CONCLUSION

It is gratifying to note the similarity between much of the ethos of the EU-driven legislation and our Health and Safety at Work Act. Nevertheless, we have to comply with all of it! Fortunately, those companies who have striven to maintain the highest standards of health and safety, and whose documentation is in good order, will find that they do not have to 're-invent the wheel'. They can comply with the majority of the new requirements with little modification to what exists already. On the other hand, for those who have not been so conscientious in these matters, there will be work to do, *but it is necessary.*

Many of the elements of the health and safety equation were formerly expressed in general terms. Now that they are spelt out much more precisely, failure to comply will not only be more obvious to inspectors, but also to one's own staff. In these litigious times, this is not the best situation to be in!

Of course, it is not just a matter of keeping out of trouble. The efficient and successful company is often the healthy and safe one – and the more progressive European companies and consortiums will be much more likely to regard safety performance as a business indicator than hitherto.

FURTHER READING

HMSO (1974), *Health and Safety at Work ACT 1974*, London: HMSO.

MANAGEMENT ACTION CHECKLIST 1

Health and Safety at Work Act 1974 (HASAWA)

Checkpoints	Action required Yes	No	Action by
Does the company hold a copy of the Health and Safety at Work Act?			
Do your training arrangements include explaining to all staff their statutory duties under HASAWA?			
Do all managers and directors understand the implications for them of Section 37 of HASAWA?			

2

ORGANIZING FOR HEALTH AND SAFETY

❖

Everyone at work has some responsibility for health and safety, but there has to be a system to ensure that these responsibilities are discharged within a sensible organizational framework. In this respect health and safety is no different from any other facet of a business.

In determining the health and safety organization needed, it is important to consider the 'staff' and 'line' resources available. This chapter examines the following two elements of the safety manpower equation:

1. *Staff* The safety officer and any other health- and safety-related specialists (e.g. occupational health nurses, hygienists).
2. *Line* Chiefly the health and safety responsibilities of line management.

STAFF

Two key references to organization and responsibility for health-and-safety are the following.

HASAWA

Section 2 of the Health and Safety at Work Act 1974 (HASAWA) covers the duties of all employers, one of which (Section 2(3)) calls for the publication of a company safety policy if there are five or more employees. This section of HASAWA also requires that the safety policy includes details of 'the *organization* and arrangements for the time being in force for carrying out the policy' [emphasis added].

There is no mention of the need for a safety officer or safety manager; indeed, ,before 1993 – apart from a few processes and some categories of construction work – there was no statutory requirement for safety officers to be appointed, although many enlightened businesses made such appointments, recognizing their importance in the drive for greater safety awareness and performance.

EU directives

Since January 1993 the position has changed radically, and there is now a clear duty to appoint safety advisers. The requirement appears in Regulation 6 of the Management of Health and Safety at Work Regulations 1992 (MHSW) (see Chapter 15). The MHSW Regulations ratify the principal EU Directive on Occupational Health and Safety, known as the 'Framework Directive' (89/391/EEC).

Regulation 6 of MHSW requires every employer to appoint a 'competent person' or persons to assist them in meeting their health and safety duties under the law. Those appointed may be employees, or external consultants, or a combination of these. Individual arrangements will vary according to the size and complexity of the business.

Official guidance on the number of persons to be appointed is, unfortunately, extremely vague. Paragraph 3 of Regulation 6 of MHSW states:

> The employer shall ensure that the number of persons appointed . . . and the time available to them to fulfil their functions and the means at their disposal are adequate having regard to the size of his undertaking, the risks to which his employees are exposed and the distribution of those risks throughout his undertaking.

COMPETENCE

The important criterion in determining who to appoint is 'competence', which is defined as sufficient training, experience, knowledge or other qualities to enable him (the safety adviser) properly to assist in undertaking the protective and preventative measures called for in occupational health and safety law and regulation. No doubt in the course of time, the guidance on competence will be augmented by recognition of specific health and safety qualifications. This will assist employers who, despite retaining responsibility for following or disregarding the advice of their health and safety experts, are obliged to appoint them without the benefit of any specific official guidance as to recognized or recommended qualifications for safety advisers.

The leading professional body in this field is the Institution of Occupational Safety and Health (IOSH), who have done a great deal to advance the case for professional standards. They should be consulted by businesses or individuals requiring further information. The address of IOSH is The Grange, Highfield Drive, Wigston, Leicester LE18 1NN (telephone: 0533 571399).

There must be full co-operation where more than one person is appointed, or where the employer opts for a combination of internal and external assistance. All persons appointed must be provided with the fullest information and assistance in order to carry out their work.

THE POSITION OF THE SAFETY OFFICER BEFORE 1993

Before considering the effects of the new requirement, it will be useful to examine the position of safety officers before 1993 – and indeed in some cases even now if their employers do not yet know the implications of Regulation 6 of MHSW! The comments that follow take account of the fact that the many variances in approach are a reflection of the size and complexity of organizations, as well as economic circumstances.

The recessions of recent years have taken their toll in terms of occupational safety appointments, although there is evidence of a new and increasing awareness of the importance of the safety professional's role. This has come about as a result of growing anxiety at the number and seriousness of workplace accidents, coupled with recognition of the fact that the sheer volume of new safety and safety-related legislation demands that it is properly reviewed and analysed if there is to be effective implementation and monitoring.

Historically the spectrum of safety organizational arrangements has extended from a safety manager with subordinate staff in large organizations at one extremity, to firms with no safety staff at all at the other! Between these extremes there will be companies who realized that they had to do something, but – chiefly for cost reasons – could not assign a person or persons solely to look after health and safety. A variation of this is where there is a wish to do everything possible, but the health and safety workload doesn't justify a full-time appointment. Another category is the company who appreciated that they had to demonstrate some concern for health and safety, or at least assign the responsibility somewhere! This was resolved by informing the administration or office manager – or in industry, the production manager or similar – that in addition to their primary role, they were now *also* safety officers!

And finally, there were cases where there was a need to provide a niche in which the long-serving employee could 'coast' towards retirement – a reward for duty faithfully performed. It is easy to visualize the scene. The production manager calls in Bill, who at 63 is slowing down. Bill is told that the company want to give him a 'lighter' job for his final two years of service, and as they feel it is about time they appointed a safety officer, Bill is just the man. To ensure that he does not become too enthusiastic, the parting comment from the production manager, 'Try not to rock the boat, Bill', was hardly a recipe for effective health and safety in the 1990s!

TIME FOR THE JOB

Time is the greatest enemy of the part-time safety officer. The more roles the incumbent has, the less attention health and safety will get. Office or administration managers, for example, will already have a portfolio covering dozens of responsibilities, and the addition of health and safety may be the last straw. Employees holding down multifaceted jobs will have inordinate difficulty fitting in time to manage safety, a role that demands time for research and planning. Frequently the other elements of a multirole job are 'real-time' in nature – mail to go out, meetings to be attended, and so on.

In these circumstances safety may be relegated to obscurity unless the individual concerned exerts a strong time-management discipline. It is essential to allocate the time needed to manage health and safety, to set realistic and achievable objectives, and to measure performance critically.

Nobody bothers about health and safety until an accident occurs – and then everyone is keen to point the finger at the part-time, untrained, inexperienced administration or office manager-cum-safety officer-cum-general dogsbody!

THE NEW CONCEPT

Since January 1993 it is simply not enough to send a newly-appointed part-time safety adviser to a one-day workshop or conference, and assume that this fits him or her for the responsibilities of the position. The most frequent criticism of such safety training by newly-appointed part-time safety officers is that a day is totally insufficient to enable them properly to understand the subject.

Since 1 January 1993 every employer has been required to publish his arrangements to cover health and safety advice and assistance; these may be to employ their own safety officers, or use external consultants, or a combination of these.

Before considering ways in which companies might arrange their health and safety advice, however, it must be stressed that on present form the EU will generate a great many more health, safety and associated directives which we will have to ratify. This means that the size and complexity of the workload for those involved directly with safety is going to be considerably increased. The reader may have concerns about the reasonableness of all this at a time of considerable financial uncertainty, but that is not a matter for this book.

HEALTH AND SAFETY ORGANIZATIONAL OPTIONS POST-1993

Although there is no specific direction as to the organizational arrangement that should be adopted, the following examples might be useful as a guide.

Small/medium-sized companies

Where there are no abnormal risks, and no employed safety officers appointed, it is likely that retention of a consultant to inspect premises at some frequency – say, about once a year – and provide an ongoing advisory service on demand, will suffice. This option should offer a cost saving compared to training an employee as safety officer.

Large structured businesses

If already employing a qualified safety officer, perhaps with subordinate staff, such organizations are likely to be self-sufficient and already meeting the new requirement.

Other businesses employing a full-time safety officer

Businesses in this category might benefit from a premises review once a year by a consultant, who might also be commissioned to provide advice/update to the employed safety officer on a regular and/or 'as required' basis. Close co-operation and rapport should exist between the two.

Businesses employing a part-time safety officer

These organizations may require consultant assistance in respect to inspections/reviews and in providing advice routinely and as required.

Self-employed persons

There is provision in the guidance on the new regulations for exemption for self-employed persons, providing that they are competent to undertake the necessary preventative and protective measures applicable to the work undertaken. Some self-employed persons may wish to take consultant advice in order to keep abreast of legislation affecting them.

It is likely that a number of consultancies will offer packages designed to provide a regular inspection and/or information service. However, whether advisers are employed or retained, it cannot be emphasized too strongly that they are *not* a substitute for good management, and their existence in no way relieves the employer, or individual managers, of their statutory health and safety responsibilities. The role of these experts is to offer advice, guidance and audit/review. It is not to assume the mantle of the line manager.

Management must understand that if their safety adviser(s)' job is to advise them, their job is to review that advice – promptly and carefully. If management opt *not* to follow the advice, they should ensure that their reasons are both supportable and well documented. In the event of a serious accident, or an inspector taking a strong line on any health and safety matter, the recommendations of the company safety adviser would be available and might be incriminating.

DUTIES OF EXTERNAL SAFETY OFFICER

Although external consultants could carry out most of the duties detailed in Table 2.1, some should be carried out by employed staff. Where no internal safety officer has been appointed, an employee could be assigned some of these duties under the guidance of the external consultant.

Where circumstances demand, some of the duties in Table 2.1 could be carried out elsewhere in the organization; for example, accident records and external accident reporting, including the external statutory reporting duty to comply with the RIDDOR Regulations, could be undertaken within the personnel department.

Table 2.1 Specimen duties of internal safety officers

- To provide a 'centre of competence' to the company in respect to occupational health and safety matters.
- To act as company interface with statutory health and safety authorities.
- To provide 'early warning' of impending and planned health and safety legislation which will affect the company.
- To advise on health and safety training needed, and where appropriate provide it.
- To act as secretary to the company safety committee.
- To carry out inspections of all company space at determined intervals.
- To ensure compliance with statutory reporting requirements (e.g. accidents), and maintain company safety statistics.
- To investigate serious accidents/incidents in liaison with nominated colleagues.
- To maintain the 'master copy' of the company safety policy, and assemble updates for ratification by the board.

OTHER KEY HEALTH AND SAFETY STAFF

Where businesses have a larger health and safety establishment (e.g. doctors, occupational health nurses, hygienists, fire officers) the roles and responsibilities of each must appear in the 'organization and responsibilities' section of the company safety policy (see Chapter 3). All of those mentioned are 'staff' within the concept of the health and safety organization.

CONCLUSION

Clearly the days of the part-time, untrained or only partly-trained safety officer are numbered, and from 1993 all businesses have to publish the arrangements that they make for safety advice, and so on. This is not simply for the benefit of visiting inspectors, but more importantly, so that every employee knows.

Furthermore, in many cases the health and safety culture will have to change, and managers and management will have to adopt a more positive stance. In the place of resentment that the safety adviser is highlighting exposures, there will have to be a complete volte-face – a desire and expectation that the safety adviser will draw attention to any shortcoming in health and safety that he becomes aware of.

LINE – THE ROLE OF MANAGEMENT

This part of the chapter covers the development of a company-wide management system for health and safety. The organization described is a model which is capable of modification/adaptation to cover a number of organizational permutations, from the single-location business to those with an HQ and numerous branches. Within this spectrum there will be firms having a number of branches of a similar kind (e.g. retail shops) and others where each location carries out a discrete part of the firm's business (e.g. warehousing, education centres, R & D, manufacturing).

LINE MANAGEMENT AND STAFF

An understanding is necessary of the line/staff differences as they apply to the management of health and safety.

The safety officer – or adviser – is the professional, and concerning his health and safety role is a 'staff' person. His job is to advise, guide and monitor, not to 'do'. It is the job of the line management to get things done – to take the decisions – and where health and safety is concerned, it is they who have clear and unambiguous responsibility in law.

The only extent to which a safety officer has 'line' responsibility in health and safety terms is in respect of his own personal behaviour as an employee (under Sections 7 and 8 of HASAWA and Regulation 12 of MHSW) and, if also a manager, for subordinates and visitors to his or her area in the same way as all other managers.

ORGANIZATIONAL PLACEMENT OF THE SAFETY OFFICER/MANAGER

In placing safety officers within the organization, the following considerations are important:

- The possibility that they may have to make unpopular, often critical, remarks and generate unfavourable reports.
- The chance that they may have to recommend stopping an operation.

The position of the safety officer within the organization must therefore be such that he or she can take that kind of action without concern for his or her future prospects!

This makes the personnel/human resources (HR) function a logical choice as the place to locate the health and safety responsibility (see Table 2.2).

HEALTH AND SAFETY AT BOARD LEVEL

Table 2.2 gives reasons why the health and safety officer should be in the personnel function, as well as making the case for assigning the strategy and support for health and safety throughout the business to the director of personnel – or human resources as this appointment is now more generally known. Where there are no directors, or none can be assigned the duties, the chief executive must assume these himself in the final analysis, and irrespective of any arrangements for delegation, it is the chief executive's responsibility anyway.

Table 2.2 Why health and safety should be located within the personnel/human resources function

- Personnel is a 'staff' function, as is health and safety.
- Personnel is a 'people'-oriented function, as is health and safety.
- In the event that the safety officer has to write reports critical of departments/functions, or to recommend stopping an operation, he or she should feel free to do so without concern for future prospects. This would be less likely to create a problem if the incumbent was in personnel.
- There is an increasing trend toward health and health-related initiatives in forward-looking organizations; such initiatives normally emanate from personnel. As 'health' and 'safety' are synonymous, it makes organizational sense to co-locate the two elements of the equation.
- The assignment of health and safety responsibility at board level suggests that the director of personnel/HR is best suited for the role. This therefore links the safety officer with the personnel director, albeit that there may be a stratum of management between the two.

The word 'responsibility' is so much a part of language that it can create misunderstanding, particularly with regard to occupational health and safety. The empirical nature of responsibility applies as much here as in every other aspect of organization. Subordinates at every level have responsibilities, but those at the top retain the ultimate responsibility. A more accurate description of the delegated task is that those doing the work have been assigned 'duties', while the chief executive retains ultimate 'responsibility'.

Therefore, it is important for the chief executive to 'monitor' the performance of the director to whom health and safety duties have been assigned. The presentation of health and safety reports and intelligence to the board provides a forum for critical analysis and review, while at the same time serving as an indicator of the performance of the director to whom the duties have been assigned. Specimen health and safety duties of a director having health and safety 'responsibility' appear in Table 2.3.

The principle of assigned duties, together with appropriate checks on performance (monitoring), are the key to effective health and safety management at each level in the organization. Moreover they are the least that a manager assigning such duties should do to demonstrate a 'reasonable' level of responsibility and care.

If problems occur, nothing will be more incriminating than the revelation that health and safety duties were assigned without subsequent regular verification that they were being carried out properly.

The list of duties in Table 2.3 is formidable. However, the advantage of having the safety officer in the same function as the director responsible, is that many board presentations on health and safety can actually be carried out by the safety officer.

Too often the enquiry into a serious accident has revealed that employees

Table 2.3 Health and safety duties of the board member assigned to health and safety

1. Foster the aims and objectives of health and safety at work throughout the company.
2. Ensure that the necessary human and financial resources are made available to carry out the company safety policy.
3. Act as chairman of the safety committee.
4. Table the following before the board:
   (a) report on health and safety status overall (once or twice per annum);
   (b) report on accident, etc., statistics (once or twice per annum);
   (c) information about proposed or impending health and safety legislation likely to affect the company (as appropriate);
   (d) details of any serious accidents/dangerous occurrences (as they occur);
   (e) details of any statutory enforcement orders issued (as they occur); and
   (f) proposed updates to the company safety policy for ratification (every other year, or more often if events demand).

perceived a danger, but were not sure what action they should take or to whom they should report. In other cases, the terms of reference of a safety-related appointment have not been clear – even to the incumbent! This is sometimes the case with the most senior (board) appointment. Directors of personnel/HR understand the rationale for assigning the duties to them; the problem is that they do not always know how to translate the general duty into specific tasks or elements. This is understandable, and it is an opportunity for the safety officer or external professional to provide guidance.

Even if the safety officer is not going to make presentations or reports himself, it will be valuable for him to attend board meetings when health and safety is on the agenda. This avoids delay in resolving problems due to the absence of professional expertise.

Obviously, all of these objectives could be achieved if the safety officer was in a different function to the director responsible, but the process would be more difficult in terms of liaison, communication and rapport.

Health and safety duties – other board members

Directors responsible for specific operations (e.g. R & D, marketing) have overall responsibility for health and safety within their functions.

GEOGRAPHIC RESPONSIBILITY

So far, this chapter has addressed appointments where the health and safety responsibility is either company-wide or function-related in nature. The third dimension concerns a span of responsibility geared to individual buildings and locations, and covering all the occupants, irrespective of the function in which they work.

In retail branches the hierarchy is clear-cut, but it may not be so in a large marketing office, in which the company houses a number of departments/functions. None the less, there must be an appointment to the position of senior or location manager, recognizing that the appointee will have other mainstream work responsibilities. The benefit of this appointment is greatest in times of emergency, or in the event that a statutory inspector wishes to speak to the senior manager in a location; for example, following a bomb warning, when a decision has to be taken about evacuating the premises, or if an inspector wishes to talk to the most senior manager, either about health and safety arrangements in general or about specific concerns. How a company responds to such requests could be important in terms of official reaction. A few words with the senior manager – and an efficient and helpful attitude and response from that manager – could avert the issue of an enforcement order or worse.

There are other non-safety reasons, of course, why there should be a clearly identified senior manager in every building. Yet the safety and health responsibility is now so important that the appointed manager's job description must refer to it – preferably as the first item!

In addition to the location senior manager there might also be a separate location safety officer – perhaps the administration or office manager – to whom the day-to-day running of safety is entrusted. If this is the case, particular note should be made of the 'competence' requirements which apply to such positions from January 1993 as a result of the Management of Health and Safety at Work Regulations (see Chapter 15).

Where the size or make-up of the workforce in a branch office does not permit the appointment of a safety officer, an employee should be appointed whose job it is to liaise with the company safety officer or external consultant. It may be that where an external consultant is used, locally appointed staff can receive some basic health and safety training to enable them to deal with routine safety matters, with the external expert providing a back-up service.

Whatever permutation is used, the arrangements must be recorded in the 'organization and responsibilities' section of the company safety policy.

CONCLUSION

The role of all managers

Every person in a position of authority is responsible for the health and safety of those reporting to them, and for others while they are within the curtilage of the manager's area of responsibility. Ideally the 'safety organization' section of the company safety policy should cover *everyone's* health and safety responsibility, but *all* managers must definitely be included, for it is managers, as the agents of their employer, who have a key role in ensuring that a healthy and safe working environment is maintained (see Chapter 3).

Employees

The responsibilities of employees have been considerably expanded by MHSW and the Personal Protective Equipment at Work Regulations (PPE) (see Chapters 15 and 19 respectively).

Finally, a clear and well-understood organization for health and safety is an important tool in creating and maintaining a healthy and safe working environment.

MANAGEMENT ACTION CHECKLIST 2

Organizing for Health and Safety

Checkpoints	Action required Yes	No	Action by
Who are the company safety advisers?			
Are they 'competent'?			
Are there sufficient advisers?			
Do all staff know who they are? (NB: Staff must have access to them.)			
Are they shown in the company safety policy organization section?			
Is there full co-operation between all safety advisers, and are they privy to all health and safety and related information, including actual and suspected hazards?			
If any safety adviser has other responsibilities: (a) is sufficient time available for health and safety; and (b) have they been properly trained and are they able to attend/receive refresher training?			
Are executive management aware of the importance of prompt consideration of safety advisers' recommendations?			
Does the board (or equivalent) receive reports similar to and at about the same frequency as shown in Table 2.3?			
In addition to central health and safety appointments, have location managers been appointed for branches?			
Do the job descriptions of location managers include their (location) health and safety responsibilities?			
Do all managers have a common paragraph in their job descriptions covering their health and safety responsibilities for subordinates, and so forth? (NB: Their failure in respect to health and safety matters could have repercussions outside the company!)			

3

THE COMPANY SAFETY POLICY

❖

Of all the duties and responsibilities of an employer under the Health and Safety at Work Act 1974 (HASAWA), producing a safety policy is often the most neglected. Yet a well-prepared safety policy can be of enormous benefit in creating and maintaining awareness of the importance of safety.

STATUTORY REQUIREMENT

Section 2(3) of HASAWA states:

> Except in such cases as may be prescribed,* it shall be the duty of every employer to prepare and as often as may be appropriate revise a written statement of his general policy with respect to the health and safety at work of his employees, and the organisation and arrangements for the time being in force for carrying out that policy, and to bring the statement and any revision of it to the notice of all his employees.

REVISION OF THE POLICY

The safety policy should ideally be reviewed once each year – every two years at most – unless any of the following circumstances demand more frequent revision:

- Change in organization and responsibilities for health and safety in the company.
- Changes in processes or operating systems.
- When changes in law or regulation necessitate.

---

*A safety policy is not required if an organization has fewer than five employees.

If the safety policy is used, as it should be, to keep staff up to date and aware of the importance which is attached to health and safety, the policy must reflect the current position. There is little point in having a policy in circulation that has become so dated by events as to be worthless. If this is allowed to happen, staff will assume that the policy is no more than a token.

At this time of dynamic change in health and safety law and regulation, the implications are for frequent update of the policy. Many of the EU-driven regulations described in Part Three call for information to be given to employees. Although this should be done in 'face-to-face' sessions, confirmation must be available in written form, and the obvious place for it is in the safety policy, probably in the form of appendices covering each specific requirement.

PHILOSOPHY AND OBJECTIVE

When an organization compiles its safety policy, it is creating a piece of law for itself. Whether the policy contains matters that could apply in a number of businesses, or that are patently exclusive to the business in question, it is important to remember that they may be checked or monitored, especially so in the event of an accident or dangerous occurrence.

It is a statutory duty to publish a safety policy, *and* to make certain that it is being complied with. Thus the policy has to be kept up to date.

If a safety-related procedure changes, and it is impractical to revise the policy at once – perhaps because other changes are expected in the short term – some formal communication or policy endorsement must be made so that everybody knows about the change. In other words, there must be a relationship with day-to-day events and the safety policy. The danger inherent in not ensuring that this happens is obvious: something goes wrong and there is an investigation by the statutory authority. Those directly involved might have been following an outdated procedure as described in the safety policy or its appendices, either because the safety policy had not been amended, or they had not been properly informed of a change in procedure.

Employees need to understand that everything in the safety policy is enforceable, not simply those things which they recognize as being part of statute law. They must not assume that policy requirements of a local parochial nature are less important.

HASAWA calls for the policy to be in writing, and that it and all revisions must be brought to the attention of all employees. It does not stipulate that all employees have to receive a copy, although most employers do provide one. Some firms require staff to sign for their copy of the policy, and others have incorporated into the receipt a certification by the employee that they will comply with it. Although it is questionable whether a certification of this kind would carry weight legally, it is undoubtedly useful in one legal sense: to prove that the employee has received a copy of the policy.

What is much more positive in the practice of issuing personal copies of the safety policy, requiring a receipt and so forth, is that staff will sense that their employer takes the whole matter of health and safety seriously. By comparison, pinning the safety policy on the company notice-board could create the opposite view – especially if a multipage document is posted with a single drawing-pin, rendering everything apart from the cover inaccessible and impossible to read!

There is no reason why the safety policy should not be incorporated in, for example, an 'employee handbook'. But this is not the ideal approach, since it relegates by implication. A policy statement tucked away at the back of a lengthy document covering the rules for, among other headings, annual leave and pensions, cannot convey a sense of the importance of health and safety, especially if the subject is simply just another section of the handbook!

IMPLICATIONS OF THE EU

The notion that the safety policy is the key motivator and reference point for health and safety in the organization seems largely to have been ignored by the EU-driven regulations described in Part Three of this book. Most of the requirements of these regulations are gradations of meaning on matters that a well-considered safety policy should have taken account of, yet these requirements are set out as additional duties, and most of them require some form of written evidence of compliance. Incredibly, the regulation that is most onerous – the Management of Health and Safety at Work (see Chapter 15) – actually has a regulation (number 4) that comes as close to repeating Section 2(3) of HASAWA in respect to safety policies as possible, without actually mentioning the words 'safety policy'!

Among the array of requirements within the new regulations, are calls for general risk assessments, display screen equipment assessments, manual handling assessments, procedures for dealing with serious and imminent danger, additional duties for employees, and so on. Most of these topics should already have been addressed by a safety policy, since that is the prime purpose of the document, namely, to consider the dangers in a business, and the measures to eliminate or ameliorate those dangers.

It is therefore consistent to incorporate the additional paperwork demanded by these new regulations into the existing safety policy, preferably as appendices to it. Not only does this make organizational sense, but also it will minimize the extent to which the safety policy has to be altered as part of the exercise involved in updating the policy to reflect the many changes brought about by the Single Market.

STATUS OF THE SAFETY POLICY

It has often been stated that the production of a safety policy, however well written, is not enough in itself. Certainly there is abundant evidence that unless the employer really believes in – and monitors – their safety policy, they have not satisfied the duty. This is why the guidance calls for the policy to be signed and dated by a company officer or board member of equivalent status.

So who will monitor the policy? What arrangements exist to ensure regular review or update? Will there be an opportunity to discuss the policy, and indeed safety generally, with all employees at least once a year? What disciplinary procedures are available to deal with breaches of the policy? All these matters are relevant and must be addressed. Moreover, someone – preferably the safety officer/manager – has to maintain the 'master copy' of the safety policy, together with all associated information (e.g. a copy of the emergency evacuation/fire arrangements).

In the event of a visit by a statutory inspector, the organization must demonstrate an efficient approach and attitude. Having a copy of all the relevant health and safety material together will be equally helpful when the subject is being reviewed.

FORMAT OF THE SAFETY POLICY

The recommended format for a safety policy is three distinct parts:

1. *Part I* The policy statement or statement of intent by the company with regard to health and safety.
2. *Part II* The organization and responsibilities to ensure that the policy is followed.
3. *Part III* Arrangements – the rules and procedures for health and safety.

PART I THE POLICY STATEMENT

This need only be a paragraph or two covering the attitude and ethos of the company board or equivalent toward health and safety; sincerity and terminology that will be meaningful and relevant to staff is important. Bland repetition of the employer's statutory health and safety duties will not appear sincere!

It would be presumptuous to suggest a precise wording for the policy statement; instead, some examples taken from existing policy statements are as follows:

> We attach the same importance to maintaining a healthy and safe working environment as we do to every other aspect of the business . . . we are concerned to provide a safe and healthy environment for our employees, visitors and others who work in our premises . . . we expect that our employees and contractors will co-operate with us and take a mature attitude to maintaining the highest standards of health and safety . . . although this is a low-risk business, we none the less take health and safety seriously . . .

PART II ORGANIZATION AND RESPONSIBILITIES

This part should show the health and safety organization from top to bottom, starting with the director responsible to the board for health and safety, and describing the health and safety responsibilities of each appointment; if appointments instead of names are used, this reduces the frequency of updating to reflect personnel changes.

For example, the HR Director is responsible to the board for health and safety in the company. His health and safety responsibilities are as follows:

- To ensure sufficient finance and human resource is available to comply with the company health and safety policy.
- To report to the board on company health and safety at least once a year, with reports as required in the event of serious accidents, the issue of enforcement orders, and so on.
- To bring proposed revisions of the company safety policy before the board for ratification every two years, or more often if circumstances demand.
- To alert the board whenever new/amended health and safety and associated legislation is proposed, if this will affect the company.
- To place company accident statistics before the board each year, or more frequently where accident rates are high.
- To support the aims and objectives of the safety policy across the company.

The responsibilities of the safety officer/manager might include:

- To provide a 'centre of competence' on all matters concerning occupational health and safety.
- To advise the director of human resources of all significant health and safety concerns/incidents.
- To maintain the 'master copy' of the safety policy, and associated documentation, and to compile the draft of subsequent updates to it for submission to the board through the director of human resources.
- To advise the director of human resources of all impending or planned health and safety regulations likely to affect the company.
- To maintain company accident records and statistics, and to ensure compliance with external reporting requirements (e.g. accidents) in accordance with the Reporting of Injuries, Diseases and Dangerous Occurrences Regulations (RIDDOR) (see Chapter 10).
- To carry out safety reviews of all space at determined intervals, and to produce reports with recommendations for improvements.
- To act as the company interface with all statutory health and safety authorities.
- To act as secretary to the company health and safety committee.

These lists are not comprehensive, and the duties can be modified to reflect the company health and safety organization.

When evaluating jobs it is important to take proper account of *specific*, as opposed to *general*, health and safety responsibilities of all managers, and, of course, to include both *specific* and *common* responsibilities as appropriate in job descriptions and job objectives, where these are used. If there are other 'specialist' health and safety staff in the company, their responsiblities should be shown. These might include doctors, occupational health nurses, chemists, hygienists, fire officers, and so on.

Part II of the policy should include the health and safety responsibilities of everyone in the company. Having covered all the 'specialist' appointments, we are left with other managers, and, of course, all employees. First, the duties of employees are shown in Sections 7 and 8 of HASAWA and Regulation 12 of MHSW (see Chapters 1 and 15 respectively). Secondly, duties for managers can be derived from the duties of employers as defined in Sections 2 and 3 of HASAWA (see Chapter 1).

It must be stressed to managers that the existence of safety professionals as described above in no way reduces or alters their line responsibility for the health and safety of their subordinates and their similar responsibility towards both visitors and contractors while they are in their designated area of responsibility.

In addition to the health and safety responsibilities of 'specialists', managers and employees generally, it might be necessary to assign 'geographic' responsibility, namely, for branch/location managers, and for any intermediate managers (e.g. regional managers or their equivalent). Where there are branches, 'geographic' responsibility is an important consideration. In the event of bomb threats or other emergencies, somebody must have overall responsibility. In a headquarters this might be the safety officer or fire officer, or their managers. In the branch office, however, there is no such infrastructure, and the senior manager/location manager has to take charge. In branches, too, allowance must be made and responsibilities allocated to cover a visit by a statutory inspector. In this event, uncertainty, or a 'laid-back' or indifferent attitude, on the part of the senior manager could make the difference between an amicable discussion and a written warning or worse!

PART 3 ARRANGEMENTS

By far the most important section of the policy, Part III should contain everything employees need to know about health and safety in their workplace. In a large, complex organization, there might be a need to produce this part locally, with Parts I and II applying across the company; given the purpose of Part I, this must of course be the same for all parts of the organization.

There are no limits to what might be included in Part III – providing that it is health- and safety-related, and that there is a need to communicate it to employees. For example, it might include: first aid, the COSHH regulations, accident near-miss and dangerous occurrence reporting, smoking, fire certificate requirements, fire and other emergency procedures, health and safety training, 'safe systems of work', breaches of the safety policy and how they are to be dealt with, and so on. Most of the subjects addressed by chapters of this book should be mentioned in this part of the safety policy.

For ease of reference and understanding there should be a contents page, if not an index; and as proposed earlier, the various assessments and so forth now demanded by EU-driven legislation could sensibly become appendices to the master copy of the policy, with a short reference to their existence in the individual (employee) edition of the policy. Given the explosion in paperwork necessary to comply with the new legislation, it would now be unmanageable to supply each employee with a copy of the policy containing the appendices.

There are some additional policy distribution requirements with reference to temporary staff and contractors. These are described in MHSW (see Chapter 15).

CONCLUSION

In the introduction to this chapter, it was stated that compliance with the requirement to produce a safety policy and keep it up to date was among the most neglected elements of workplace health and safety law.

This is unfortunate not only because it breaches the legal duty, but also because it obliges employers to develop other ways to get the health and safety message to their staff. In the same way that each new health and safety regulation is introduced under the authority of the Health and Safety at Work Act, so each change or improvement to health and safety in the business could be introduced under the authority of the company safety policy.

Two benefits accrue. First, employees would be encouraged to take an interest in what is in the policy, and would thus become accustomed to its authority and value as a mirror of the current statutory position. Secondly, when the statutory inspector visits, their first request could be to see the safety policy. If this request meets with a positive response – for example, prompt display of an up-to-date document, with evidence of ongoing review and management commitment, the inspection may end there. However, all too often such a request is met with embarrassment and confusion. Various excuses are offered for being unable to produce a document which ought to be readily available and to be seen on every manager's bookshelf. When chaos and confusion follows such a request, it is giving negative signals to the inspector, signals which suggest that a few more stones need to be turned over . . . !

FURTHER READING

Health and Safety Executive (1987), *Writing a Safety Policy Statement: Advice to employers*, HSC no. 6, London: HSE Books. Available free from HSE enquiry points (telephone: 0742 892345).

Health and Safety Executive (1989) (rev. edn), *Our Health and Safety Policy Statement: Guide to preparing a safety policy statement for a small business,* London: HSE Books. Priced at £2.50.

MANAGEMENT ACTION CHECKLIST 3

Company Safety Policy

Checkpoints	Action required Yes	No	Action by
Has the current edition of safety policy been published within the last two years?			
Is it signed and dated by the managing director or a board-level director?			
Do regular update/review procedures exist?			
Have the EU-driven regulations been included?			
Is COSHH mentioned in Part III of the company's safety policy? It should be (see Chapter 10).			
Does every employee receive a copy?			
If yes, are new starters automatically issued with a copy? At induction?			
Does every employee receive 'face-to-face' instruction/briefing on the policy?			
Is the policy arranged to show the three parts described in this chapter – or similar?			
Are the specific duties of key appointment holders and other managers and employees shown?			
Do employees understand that their duties are laid down in statute/regulations, and that breaches are criminal offences?			
If the policy is incorporated into an 'employee handbook', is it accorded appropriate status? For example, does the cover emphasize that company safety policy is inside, and is the policy properly separated and highlighted?			

4

HEALTH AND SAFETY MONITORING

❖

This chapter is probably the most important in this book; its subject matter is not specific to health and safety, but is relevant to the entire spectrum of business.

The life-blood of business is intelligence – information on every aspect of a company's operations, as well as that of its competitors! No business can survive without monitoring and feedback. In the age of the computer, production of feedback has been brought to the acme of perfection, and a computer-generated report can be produced for virtually anything.

However, health and safety reporting and feedback has not shared in the reporting increase, and with a few exceptions such reporting and feedback as does occur is based upon precepts established many generations ago. Inspections still concentrate on poor housekeeping, failure to wear prescribed personal protective clothing and equipment, blocked fire escapes, and so on, without considering why these things happen. It is a case of highlighting the effect without considering the cause.

This superficial approach to monitoring probably has its roots in the days when supervisors in industry were expected to walk through their departments once a week and highlight breaches of the kind described.

In many companies the supervisor's weekly walkabout was not so much a matter of resolving problems, but rather that supervisors vied with each other to see how many infringements of rules and regulations they could find. The fact that the same safety violations arose week after week, and often perpetrated by the same employees, was incidental!

Although matters are improving, there are still many companies where this 'superficial' policing continues. The focus of attention is the immediate breach, without considering why it happens, and what management action and/or improvement in existing systems and procedures is necessary to correct the situation once and for all.

Before 1993, effective health and safety monitoring and feedback was the hallmark of the efficient company – a matter of good business practice. From January 1993, however, effective monitoring arrangements became compulsory by virtue of Regulation 4 of the Management of Health and Safety at Work Regulations 1992 (MHSW) (see Chapter 15). Regulation 4(1) requires every employer to make and give effect to such arrangements as are appropriate, having regard to the nature of his activities and size of the business, for the effective planning, organization, control, *monitoring and review* of the various health and safety measures which are in place. There can now be no excuse for not having a monitoring system in operation in addition to monitoring for specific elements of the health and safety equation which are demanded by law.

SELF-CERTIFICATION

The introduction of managers' health and safety self-audit certification systems has been instrumental in raising the health and safety profile generally, but its real value has been in making managers realize that the primary responsibility for health and safety is theirs: they 'own' the problem.

It is surprising that it has taken so long for companies to adopt the self-certification approach with health and safety, given that it has been integral to so many other aspects of business for years.

Some environmental health departments in local authorities have seized upon the self-certification procedure by asking companies to complete a questionnaire relating to their compliance with various aspects of health and safety law and regulation. They point out in the covering letter sent with the questionnaire that where they feel the responses indicate a less than satisfactory position, they will visit the premises.

This is an interesting development, which must leave employers who have been indifferent or dilatory in these matters in a difficult position. It is easy to imagine the concern which many employers must feel at receiving an invitation to tell all, particularly where there is very little to tell!

Self-certification is an excellent method of monitoring but, like all information, it needs to be checked. Unless managers know that their answers could be carefully scrutinized, there will inevitably be some 'white lies' included.

Also, if a self-certification programme is embarked upon, the way that questions are put can make all the difference. Badly framed questions can result in a feeling of comfort which is misplaced. For example, the question 'Do all your staff understand the COSHH regulations?' could be rephrased as 'What was the method you employed to familiarize your staff with the COSHH regulations?' The second version of the question is of more benefit to both the responding manager *and* the company in the following ways.

The manager The manager has not been 'led' by the question and cannot glibly answer 'Yes' in the hope and belief that he will not be contradicted. He must think about the method he used, which might have been no more than

putting a notice on the department notice-board.
The company It will not only find out how many managers have and have not explained COSHH to their people, but it will have gained valuable feedback on the approach taken by managers to discharge their duty to train.

When these answers are linked to those to other questions, patterns of management behaviour emerge which will be of value to training departments, and will be important when considering future communications as well as the strategy for the health and safety programme. And many questions can be framed in a way that provides valuable feedback without creating a 'Big Brother' image.

However, in the example question above about COSHH, another way to put this question would have been to offer options: 'What was the method you used to familiarize your staff with the COSHH regulations? Was it: department meeting; face-to-face session with each employee individually; information placed on department notice-board; information in written form given to each employee with invitation to discuss if they had problems; or other? Please describe.' This is still a 'leading' approach, and therefore wrong, because it offers the uncommitted manager an opportunity to respond without thinking too consciously about what he actually did – or perhaps did *not* do!

The frequency of the self-certification exercise will be a function of the nature of the business. Those with high-risk operations might require certification every six months, while the commercial sector would feel that annual certification was appropriate.

Managers should not be allowed to become blasé about the self-certification programme, or frightened by it. They will surely become indifferent if they believe that it is just an exercise which gives them a lot of work to complete, with no real benefit deriving from it; while at the other extreme some might be concerned that by certifying that things are as described in their responses to questions, they will be 'liable'.

If the self-certification programme is to bring real benefits in health and safety terms, both extremes of approach must be taken into account. The blasé manager's mind will be concentrated sharply if he realizes that his responses may be verified from time to time, or if the safety adviser contacts him to ask for further or better details of particular responses. As the purpose of the exercise is to ensure that health and safety is being properly managed, questioning of managers is crucial to the process.

In the case of managers who might be concerned, either because they feel that certification commits them – which, of course, it does – or because they are uncomfortable at their lack of knowledge of the subject to which a particular question refers, there should be a preamble to self-certification questionnaires that should be both a warning and a comfort. First, it should explain that a manager should not give an affirmative reply to *any* question unless it can be substantiated. Secondly, the statement should then offer expert advice to the manager, in the event that he or she has any difficulty in understanding the question, through a telephone advice number – the advice to be available anonymously if so required.

OTHER HEALTH AND SAFETY MONITORING

STATUTORY MONITORING AND FEEDBACK

Although this chapter opened by claiming that health and safety monitoring and feedback has made slow progress, it has not been moribund. In a number of respects elements of health and safety have been the subject of reporting requirements for many years. For example: reporting of injuries, diseases and dangerous occurrences under RIDDOR (see Chapter 9); reporting of fire precautions checks and fire drills to comply with the conditions of a fire certificate (see Chapter 11); inspections of pressure vessels, lifts, fork-lift trucks, and so on.

Of these statutory reporting duties, the reporting requirements under the RIDDOR regulations is the only example of reactive monitoring in that it deals with situations which the health and safety arrangements in place have not prevented. The other statutory reporting requirements are proactive in that they are concerned with making sure that all is as it should be, which is also the objective of the various monitoring measures described in this paragraph.

AUDITING, INSPECTION, REVIEW

The health and safety inspection review process has generated a language and meaning all of its own. Never has one primary activity – that of finding out how a department or company are performing in health and safety terms – been given so many different names, with particular meanings ascribed to some of them that would as well fit any of the others. There are safety audits, surveys, inspections, tours, sampling, reviews, checks, spot checks, and many others.

Whatever the purist may say, the fact is that a company can call these activities whatever they like. The important point is that there should be a clearly defined system of monitoring for health and safety which is understood across the organization, and the objective of which is to highlight procedures that are wrong or need improving. For larger companies the system might require checks at a number of levels, in smaller ones less.

No matter who carries out the check or review, the following principles apply:

1. The product of the review (or check, audit, and so on) should be a report that defines specific problems and breaches of safety, and where these breaches appear to be symptomatic of the failure of management and/or management systems, the report should say so.
2. The manager next above the manager whose department or function is the subject of the review, should receive a copy of the report.
3. The manager whose operation is the subject of the report should respond to the head of the review group, with a copy to his or her own

manager, describing the plan and timetable which they have established for the resolution of the matters raised in the report. There may be a variation of this where the reviewing team have themselves set deadlines for resolution of some of the issues due to their seriousness.

Health and safety reviews and so forth are *not* an alternative to the managers' self-certification programme which has been described; rather they are complementary to it. Indeed, they are an important check on the responses of managers to the self-certification exercise. Where a department conducts its own review, this might be to confirm to the department manager completing a self-certification that the situation is as it should be.

INTERNAL AUDIT

There is growing awareness of the important contribution which an internal audit department can make in terms of monitoring health and safety performance. Internal audit functions are of their nature required to visit all parts of the company, and the inclusion of health and safety in their checks is not only economic in resource terms, but is tangible evidence of the fact that health and safety is integral to business, not an appendage to it.

Although a higher than average level of knowledge of health and safety is useful to internal audit department staff, it is not essential, and absence of detailed knowledge should not prevent this important resource from playing its part in the monitoring effort. If internal auditors work to a checklist drawn up by the company safety adviser, the questions will be those most helpful in establishing the health and safety status within the company.

The involvement of internal audit in relation to health and safety monitoring should be communicated to all managers. This will provide the important motivation necessary for managers to take the self-certification assessment exercise seriously.

SAFETY REPRESENTATIVES

The role of safety representatives in a company with recognized trade unions (see Chapter 6) includes carrying out inspections of the workplace at specified intervals and when there are serious accidents or when work processes change. These inspections provide management with valuable information from a different perspective.

THE REVIEW PROCESS

There are a number of different possible approaches to the review process, and the method selected for different types and levels of review may vary.

At department level, it is possible to work to a predetermined pattern by considering each item and verifying that it meets the standard. On the one hand, this might be too time-consuming in the event that one person is carrying out a review of a complete building. On the other hand, where a team is reviewing a building, team members could be tasked to review an element or group of elements. A more informed view can be obtained if the composition of a team includes different disciplines (e.g. chemists, hygienists, safety experts).

CONCLUSION

Now that management are required by MHSW to develop effective monitoring and control systems where these do not exist already, there is no reason why health and safety cannot be fully integrated into computerized reporting systems, which will further support the principle that this subject is no different from any other in the business spectrum. Unless a company has a properly conceived and operated health and safety reporting system, it can never with confidence assert that it is discharging its health and safety responsibilities under the law.

MANAGEMENT ACTION CHECKLIST 4
Health and Safety Monitoring

Checkpoints	Action required Yes	No	Action by
Do you have a managers' health and safety self-certification programme in place?			
If yes, what analysis/verification of responses takes place?			
Do your internal audit department include health and safety in their work?			
If yes, do they liaise with your company safety adviser on the matters audited?			
If yes, are all managers aware that their health and safety performance, including their completed self-certifications, may be checked by internal audit?			
What inspections/reviews take place in your company?			
Are these organized properly and are the terms of reference clear?			
Are the managers of those being reviewed copied on the audit/inspection reports?			
Are managers of departments reviewed required to formally respond to inspection reports describing the action programme with dates?			

5

INSPECTION, PROSECUTION AND SANCTION

❖

All employers should understand what the powers of statutory inspectors are, and what sanctions are available to the courts to deal with breaches of health and safety law and regulations. This is in no sense a negative approach – rather a matter of understanding all the elements of the health and safety equation.

Although there is now a formidable array of sanctions available to the courts to deal with those in breach of health and safety law, factory inspectors and environmental health officers are encouraged to bring about compliance and improved standards by persuasion and logic rather than resorting to the courts; unfortunately cases still occur where nothing short of court direction will bring about improvements. This is a sad reflection on the attitude of the employers concerned, who, by their indifference or disregard for health and safety, are implying that employees are unimportant – an attitude reminiscent of the nineteenth rather than the twentieth century!

It might be argued that the over-abundance of law is itself the cause of much non-compliance: that it is not a matter of wilfulness, but rather of ignorance. This view is not supported by the evidence: over a third of health and safety prosecutions relate to breaches of the primary legislation (HASAWA), and not to regulations dealing with specific matters. This suggests that those charged were not even prepared to comply with the fundamentals, far less the detail.

Frequently, when breaches of safety law are the subject of prosecution, there is evidence of considerable efforts by inspectors to get recalcitrant employers to improve their ways, and prosecution has been initiated only as a last resort. Given the widespread concern in the country at the huge burdens placed upon businesses by the six sets of EU-driven health and safety regulations, the probability is that inspectors will not resort to the courts for non-compliance with these regulations, particularly where some attempt is being made to comply – and where it is possible that the regulations could be modified as a consequence of the review of all EU-driven legislation which is currently taking place.

However, inspectors may well take a tougher line where there are breaches of pre-1993 legislation, on the basis that if employers cannot achieve compliance with earlier legislation, they will never catch up with that deriving from the EU.

WHEN THE INSPECTOR CALLS . . .

It is vital to ensure that visiting inspectors are received and accompanied by staff who have both authority and knowledge. To entrust visiting inspectors to junior staff who clearly cannot respond meaningfully to all the questions is to invite further visits or correspondence – and certainly some criticism. If the safety adviser appointed in accordance with Regulation 6 of MHSW (see Chapter 15) is an employee, it makes sense for him or her to interface with inspectors.

All necessary documentary information and evidence must be readily available at the time of the inspection or visit. This should include at least the following items:

- Master copy of the company health and safety policy.
- Fire certificate and fire log-book.
- Records of statutory inspections (e.g. for lifts, cranes, pressure vessels).
- Health and safety training records.
- Assessments (e.g. COSSH, DSE, MHO, general risks).
- Procedures for serious and imminent danger.

POWERS OF INSPECTORS

Table 5.1 shows that inspectors have been invested with considerable authority. Inspectors can themselves prosecute a case under HASAWA or its subordinate regulations in magistrates' courts.

ENFORCEMENT ORDERS

Enforcement orders were introduced by HASAWA and provide inspectors with an important incentive where employers are wilfully negligent or are not prepared to make necessary health and safety improvements. There are two kinds of notices as they are styled.

Improvement notices

Improvement notices are issued where an inspector is of the opinion that a person is contravening one or more of the relevant statutory provisions.

The order directs an employer to carry out prescribed actions or improvements within a specified period, generally about 28 days. An improvement notice can be appealed within 14 days of receipt, the appeal being heard by an industrial tribunal. If an appeal is lodged, and pending the outcome,

Table 5.1 Powers of inspectors

- To enter premises at reasonable times or at any time where there is danger.
- To take a policeman with him if he is of the opinion that he will be obstructed.
- To take with him any other persons duly authorized by his enforcing authority and to take any equipment or material required for any purpose.
- To carry out an inspection or investigation.
- To require parts of premises to be left undisturbed as long as is deemed necessary.
- To take measurements and photographs and recordings as necessary.
- To take samples of articles or substances found in the premises, or to sample the atmosphere in the vicinity of the premises concerned.
- To have any article or substance dismantled or subjected to any process or test.
- To take possession of an article or substance that is necessary to facilitate an examination, to prevent tampering or to make available as evidence in proceedings.
- To require a person to give information and to make statements.
- To inspect and take copies of relevant documents or registers.
- To require assistance as necessary.
- Any other power which is necessary.

no further action need be taken on the improvement notice.

However, because an improvement notice is the less serious of the two enforcement orders, there is a tendency for some employers to treat them as non-urgent matters. This tendency should be resisted. Sometimes inspectors have returned to a premises on which they have served an improvement notice, only to find that the employer has not only taken no action to comply with the requirements of the notice, but has actually mislaid the notice itself. This easily occurs in large organizations where the recipient passes the document to whoever they believe should take action, but the action addressee might be on leave or on a course, and the document then starts doing the 'in-tray' rounds.

Therefore the legal department, or another department such as internal audit, should be made responsible for monitoring every enforcement order that arrives to ensure (a) that action is taken, and (b) that it is taken within the timescale called for in the notice. It may be that the recipient intends to comply with the notice, but cannot do so within the timescale allowed – perhaps due to non-availability of parts and so forth. In these cases a telephone call to the inspector requesting an extension of time will usually result in this being granted.

Prohibition notices

Prohibition notices are clearly more serious than improvement notices, and may be issued by an inspector if he is of the opinion that an activity may give rise to the risk of serious personal injury. It does not require the operation in question to be in specific breach of a relevant statutory provision; the sole criterion is the risk of injury.

Prohibition means what it says – the operation or machine must stop until the required improvements have been made. Furthermore, there is nothing to prevent a summons being served concurrent with a prohibition order for the same breach.

It is possible to appeal a prohibition notice, but the notice must be complied with pending the hearing, which is also before an industrial tribunal. Unless, therefore, a company served with an improvement or prohibition notice are convinced that what they are doing is not dangerous, and is not offending against a regulation, they are advised to comply with the notice. These notices are not served lightly or unless the inspectors are satisfied that they will be upheld by industrial tribunals in the event of an appeal.

SANCTIONS

When HASAWA first appeared, the expression 'I don't want to go to prison for a safety offence' was sometimes heard. This was because HASAWA introduced imprisonment as a sanction for a health and safety offence for the first time.

In fact, at present the chances of being imprisoned are extremely slim – almost non-existent on past form. However, increased powers being given to magistrates, which are described later, may change this. Moreover, with growing concern at the number and severity of industrial injuries, the chances of custodial sentences must increase. Tables 5.2 and 5.3 detail the offences specified under HASAWA which could attract imprisonment or a fine, or both.

POWERS OF THE COURTS

From 1974 until 1992, the powers of courts in health and safety matters were as follows:

1. *Magistrates' courts (and sheriff courts in Scotland)* Fines up to £2,000.
2. *Higher courts* Unlimited fines and up to two years' imprisonment.

From 1992, however, magistrates' courts, and in Scotland sheriff courts, have had their powers increased to fine up to £20,000 for most health and safety offences, plus the power to sentence to imprisonment for up to six months for offences relating to breaches of enforcement orders (improvement and prohibition notices). These are dramatic increases in the powers of magistrates, but it remains to be seen whether they are used as intended.

One of the problems experienced by the HSE, particularly with regard to

Table 5.2 Offences under the Health and Safety at Work Act 1974

- Failure to comply with the general duties (Sections 2–7). These are the sections covering the duties of employers, employees, persons in charge of premises, manufacturers, suppliers, and so on.
- To contravene Sections 8 or 9. These are the duties not to interfere with safety equipment, and the employer's duty not to charge for safety clothing and equipment, or for necessary medical examinations respectively.
- To contravene health and safety regulations.
- To make a false statement.
- To make a false entry in a register.
- To obstruct an inspector.
- To prevent any persons appearing before an inspector.
- To pretend to be an inspector.
- To contravene the terms of a licence.
- To contravene a notice.
- To use or disclose information improperly.

Table 5.3 Offences liable to be punished by imprisonment

- Carrying out activities without a licence, where a licence is required.
- Contravention of a term or condition or a restriction of a licence.
- Offences involving explosives, such as attempting to acquire, possess or use.
- Contravention of a prohibition notice or a requirement imposed by a notice.
- Contraventions in respect of using or disclosing information.

health and safety cases heard by magistrates, is the derisory fines imposed. It was not uncommon to read sensational headlines reporting notional fines of a few hundred pounds in cases where victims of workplace accidents had lost a limb or an eye, although of course much greater sums would be involved in the civil actions for compensation and damages. In higher courts there is not such a problem, and fines of up to £750,000 have been imposed. And there are likely to be more of the same, as public awareness and concern at the number and seriousness of health- and safety-related offences grows.

The Health and Safety at Work Act is covered in Chapter 1 and, in relation to the personal responsibilities of directors and others in authority, mentions a recent case where a company director and his firm were each fined for an offence, the company under Section 33 and the director under Section 37 of

HASAWA; the director was also disqualified from holding the office of director for a period of two years under the Company Directors Disqualification Act 1986. This was an unusual twist to sanctions for health and safety offences, and may indicate an approach that will gain favour where directors and senior executives are found wanting in future.

Nevertheless, given the level of fines imposed for health and safety offences since 1974, it has to be said that the outcome for most companies convicted has not been damaging in financial terms. What is far more damaging is the effect upon the reputation or image of the convicted company, which probably had a far more salutary effect that the fines and costs of appearing in court.

FURTHER READING

HMSO (1974), *Health and Safety at Work Act 1974*, London: HMSO.

MANAGEMENT ACTION CHECKLIST 5

Inspection, Prosecution and Sanction

Checkpoints	Action required Yes	No	Action by
Is there a standard procedure for dealing with visits to company premises by an enforcement authority inspector?			
Does the procedure call for the assembly in one file of originals or copies of the fire certificate, safety policy and other material (e.g. assessments)?			
Is there a procedure which requires a manager who receives an improvement or a prohibition order to immediately telephone your legal department or some other function charged with monitoring enforcement order compliance?			

6

SAFETY REPRESENTATIVES AND SAFETY COMMITTEES

❖

There can be no doubt that employee involvement and consultation with respect to workplace health and safety offers considerable benefits to management and workers alike. Accidents can be reduced, morale enhanced, costs contained and industrial relations improved if there is regular and effective consultation on health and safety issues.

For historical reasons this rationale has not found widespread favour in Britain's businesses, and there has been a marked absence of positive legislation – apart from that pertaining to unionized firms – to address the problem. Given the positive approach of the EU with regard to employee participation, which is reflected in all the directives relating to workplace health and safety, our ratifying regulations, with some exceptions, appear weak and vacillating.

This chapter covers the background to this subject, describes the legislation that exists, and concludes with a section covering the establishment, composition and operation of a safety committee in an organization which has no trade unions.

BACKGROUND

THE ROBENS COMMITTEE

The Robens Committee, set up to examine the state of occupational health and safety in Britain, and whose report was the precursor to the Health and Safety at Work Act 1974, set considerable store by the appointment of safety representatives and effective safety committees. Extracts from that committee's report in July 1972 stated:

> Most of the employers, inspectors, trade unionists and others with whom we discussed the subject are in no doubt about the importance of bringing workpeople more directly into the actual work of self-inspection and self-

regulation by the individual firm. There is no real dispute about these aims. We are left with the question – can legislation help, and if so how?

We believe that the best answer would be a statutory requirement dealing in general terms with arrangements for participation by employees. Our view is that the involvement of employees on safety and health measures is too important for new occupational safety legislation to remain entirely silent on the matter.

We recommend therefore, that there should be a statutory duty on every employer to consult with his employees or their representatives at the workplace on measures for promoting safety and health at work, and to provide arrangements for the participation of employees in the development of such measures. The form and manner of such consultation and participation would not be specified in detail, so as to provide the flexibility needed to suit a wide variety of particular circumstances and to avoid prejudicing satisfactory existing arrangements.

Two years later, when HASAWA became law, the only explicit reference to consultation (Sections 2(4)–2(7)) applied to those firms that had recognized trade unions. In 1977 the arrangements for safety representatives and safety committees in workplaces where recognized trade unions existed were promulgated by the Safety Representatives and Safety Committees Regulations 1977.

THE HEALTH AND SAFETY COMMISSION AND EXECUTIVE (HSC/HSE)

It was not until 1976 that the HSC first published a short pamphlet titled *Safety Committees: Guidance to employers whose employees are not members of recognised independent trade unions.* The intention of this pamphlet was to remind employers without accredited trade unions that consultation with their workers on health and safety matters, if not expressly required by law, is good practice, and ought to be followed!

Given the importance which the Robens Committee attached to worker consultation and participation, it is surprising that the HSC should take two years after the passing of HASAWA to provide written guidance to non-union employers on safety committees.

Even more surprising is our reaction to the EU pronouncements on worker participation described next.

THE EUROPEAN UNION (EU)

The ethos of all EU direction on workplace health and safety is that of consultation – 'balanced participation' is a frequently used term.

Since January 1993 we have had to comply with a number of European directives concerned with occupational health and safety. These were introduced under the aegis of the EC Framework Directive (89/391/EEC) which is covered in detail in Part Two. The preamble to this directive refers specifically to worker participation:

> whereas they [workers] must also be in a position to contribute, by means of balanced participation in accordance with national laws and/or practices, to seeing that the necessary protective measures are taken.

Article 11 of the directive deals solely with such participation, and is titled 'Consultation and participation of workers'. Paragraphs 1 and 2 of Article 11 are most apposite and include the following:

> Employers shall consult workers and/or their representatives and allow them to take part in discussions on all questions relating to safety and health at work.

Thus we have the following principles:

- Consultation with workers.
- The right of workers and/or their representatives to make proposals.
- Balanced participation in accordance with national laws and/or practices.

Article 11 also states that workers or workers' representatives with specific responsibility for the health and safety of workers shall take part in a balanced way, in accordance with national laws and/or practices, or shall be consulted in advance and in good time by the employer with regard to:

- Measures which may substantially affect health and safety.
- The designation of workers to be nominated to assist the employer in carrying out his health and safety responsibilities.
- The employer's 'risk assessment' relating to risks at the workplace and the measures he proposes to mitigate/eliminate such risks.
- The consultant(s) he proposes to use, in the event of the employer deciding to enlist 'outside help' (i.e. consultants) to assist him in discharging his health and safety responsibilities.
- Planning and execution of health and safety training.

Some of these consultation requirements are faithfully translated in our subordinate regulations made to comply with EU directives covering specific matters; for example, operation of display screen equipment (DSE) (see Chapter 16). It is only possible to comply with the Health and Safety (Display Screen Equipment) Regulations if every DSE 'user' is consulted, and participates in conclusions about his or her workstation.

However, many of the matters on which the EU expressly require consultation are not reflected in our ratifying regulations, or if they are, it is most unlikely, given the *modus operandi* of many companies, that employees would be consulted.

Perhaps the best example to make this point is the requirement for every employer to appoint one or more persons to assist him in discharging his health and safety responsibilities – Regulation 6 of MHSW (see Chapter 15). In a non-union company it is unlikely that the employer would consult with staff about these appointments, even if it was proposed to make them from within – still less so if it was decided to retain a consultant for the purpose.

Given the strong emphasis on worker participation in all these matters, it might be thought that in the UK the requirement to recognize safety representatives and establish safety committees – which is currently only applicable to unionized concerns – would be extended to cover *all* businesses. Not so! Instead, only the Safety Representatives and Safety Committees Regulations 1977 have been amended to strengthen the position of safety representatives in unionized concerns. It is difficult to see how the EU requirements for consultation and participation will be satisfied by these proposals. Presumably, in cases where workers have no nominated representatives, their managers will decide what is best – hardly in the spirit or letter of the EU directives!

Although this summary of the legislative position appears bleak, nevertheless it is only one side of the picture. A great many non-union companies have established safety committees and/or appointed safety representatives. They have done so because they recognize the important part that they can play in the health and safety equation. Others may have formed safety committees because they do not wish their employees to feel 'left out': that is, more exposed than their counterparts in companies which have recognized trade unions.

Whatever the motivation, there will be considerable benefits – not only of a practical kind, but in terms of employee morale and attitude – in drawing employees into the health and safety process in a positive way.

SAFETY REPRESENTATIVES AND COMMITTEES IN UNIONIZED COMPANIES

In this context it will be useful to examine the Safety Representatives and Safety Committees Regulations 1977, which set out the functions of safety committees and the role of safety representatives. Although applicable only to companies with recognized trade unions, these regulations none the less provide an excellent reference point for the non-union company, and serve as a model for those who have formed or are contemplating the establishment of a safety committee, or intend appointing/recruiting safety representatives.

THE SAFETY REPRESENTATIVES AND SAFETY COMMITTEES REGULATIONS 1977

Scope

These regulations are binding upon companies having 'recognized trade unions': that is, an independent trade union as defined in Section 30(1) of the Trades Union and Labour Relations Act 1974.

Appointment of representatives

The regulations allow recognized trade unions to appoint safety representatives from among the employer's workforce where one or more of their members are employed. The appointment must be in writing.

Normally the appointed representative(s) must have been in the employ of the company concerned for two years – or must have had two years' experience of similar work. Exceptions may be where the employer or the workplace is newly established, where the work is of short duration, or where there is a high labour turnover. The representative ceases to hold the office in the following circumstances:

1. The appointing trade union write to the employer saying so.
2. He or she ceases to be employed by the company.
3. He or she resigns.

There is no regulation relating to the numbers of safety representatives that may be appointed. Ideally the number should be agreed in consultation between the employer and the union(s) concerned, taking account of the numbers employed, the level of risk inherent in the work, whether or not there is shift working, and so on.

Liability of safety representatives

Although safety representatives are allowed to undertake a number of functions, which their employers must permit, they are not legally obliged to carry out any or all of them, nor are they criminally liable in respect to the performance of these functions.

These representatives have the same legal duties as other employees in accordance with Sections 7 and 8 of HASAWA: to look after their own health and safety and that of others who may be affected by their acts or omissions; to co-operate with their employer in all the measures he takes to discharge his statutory duty with regard to health and safety; and not to interfere with or misuse anything provided by the employer in the interests of health and safety.

The above duties have been substantially increased from January 1993 by Regulation 12 ('Employees' duties') of the Management of Health and Safety at Work Regulations 1992 (MHSW) (see Chapter 15).

Functions of safety representatives

General To represent employees in consultation with the employer, and to ensure the health, safety and welfare of workers by promoting, developing and monitoring appropriate measures.

Specific The specific functions of safety representatives are as follows:

1. To investigate potential hazards and dangerous occurrences at the workplace – whether or not these are drawn to their attention by the workers they represent – and to examine the causes of accidents at the workplace.
2. To investigate complaints by any employee represented relating to that employee's health, safety or welfare at work.
3. To make representations to the employer on matters arising from the above or in respect to any other general health, safety and welfare matters pertaining to employees in the workplace.
4. To carry out inspections of the workplace.
5. To represent employees for whom he is appointed in consultations at

the workplace with inspectors of the HSE and any other enforcing authority, and to receive information from such inspectors in accordance with Section 28(8) of HASAWA.

6. To attend meetings of safety committees.

Provisions for time off An employer must permit a safety representative time off with pay during the employer's normal working hours for the following purposes:

1. To permit the safety representative to perform the functions described above.
2. To undergo reasonable training for the performance of his duties.

Training In most cases the safety representative's own union, or the TUC, may provide or recommend to the employer the appropriate training, but this is not binding upon the employer, who in any event may require that, where a number of safety representatives are employed, their attendance for training is staggered to take account of the work arrangements. A safety representative may appeal to an industrial tribunal if his employer fails to permit time off in accordance with Regulation 4(2).

It has been held, however, that where an employer offers suitable alternative training, this may satisfy the requirement, since the duty in Regulation 4(2) is to allow time off for training which is reasonable in all the circumstances (*White* v. *Pressed Steel Fisher Ltd* [1980] IRLR 176).

Inspections Safety representatives may inspect the premises where their members work as follows:

1. At least once every three months, or more often if the employer agrees. The employer is to be informed beforehand in writing.
2. Where there has been a substantial change in working conditions, or when new information from the HSC/HSE relevant to hazards since the last inspection has been received, even if three months have not elapsed.
3. Where there has been a notifiable accident or dangerous occurrence as defined under RIDDOR (see Chapter 9). Where reasonably practicable, the employer should be given warning, and should provide such facilities as the safety representative may reasonably require (e.g. independent investigation, facilities for private discussion with witnesses). The employer's representative may be present during these inspections.

Inspection of documents Safety representatives, in the course of performing their duties under Section 2(4) of HASAWA or under these Regulations, have the right, provided they have given the employer reasonable notice, to inspect and take copies of any document relevant to the cause of their enquiry, subject to the following limitations:

- National security is involved.

- A statutory prohibition is involved.
- An individual can be identified – unless the individual agrees.
- The information is not relevant to health, safety or welfare.
- The information was acquired by the employer for the purpose of bringing, prosecuting or defending any legal proceedings.

Safety committees

1. A safety committee must be established by an employer within three months of being requested to do so in writing by at least two recognized safety representatives.
2. The following procedure shall be followed when an employer is requested by safety representatives to form a safety committee:
   (a) the employer shall consult with the safety representatives who made the request, and with any other safety representatives representing workers who will be affected by the proposed committee;
   (b) the employer must post a prominent notice stating the composition of the committee and the workplaces covered by it; and
   (c) the committee shall be established within three months of the request to do so being received by the employer.

The detailed arrangements to comply with these provisions should be settled by joint negotiation, together with the terms of reference of the committee. Once agreed these should be included as an appendix to the company safety policy.
Guidelines The guidelines for safety committees are as follows:

- Size to be as compact as possible, subject to adequate representation of the interests of both management and employees.
- Management numbers not to exceed employee representation.
- Management membership should include all relevant areas (e.g. personnel, works engineering).
- Specialists should be *ex officio* (e.g. safety officer, medical officer, industrial hygienist).
- Other specialists to be co-opted as the need arises.

Safety representatives do not automatically qualify for membership of the safety committee.

Meetings should be planned well in advance, with the frequency determined by local circumstances, nature of the enterprise, and so on. Minutes should be published promptly, and should be circulated to company executives as well as all committee members.

SAFETY REPRESENTATIVES AND COMMITTEES IN NON-UNION FIRMS

Although the Safety Representatives and Safety Committees Regulations 1977 provide a useful reference for the non-union company, there must be differences of approach to take account of organizational and other variances. For example, there is no recognized agency for training safety representatives and committee members in the non-union field. This section discusses some of these differences.

CHOICE AND TRAINING OF SAFETY PERSONNEL

Trade union safety representatives can take advantage of the excellent training provided either by their own union or the TUC. Indeed, those who have received such training frequently return to their workplaces better informed about workplace health and safety than their employers' safety staff!

Of all the differences in approach, training is perhaps the most important of all. Although the start of the Single Market necessitates that *all* those concerned with workplace health and safety are properly trained, experience shows that for a great number of companies in the UK, the position of safety officer has frequently been added to the incumbent's primary job; office or administration managers are often compromised in this way. The appointment is often the result of a safety emergency of some kind, or official intervention which has served as a catalyst to remind senior management that they ought to have some expertise at hand. Unfortunately this dilemma can result in a precipitate and illogical choice of personnel. The hapless incumbent of the newly-created part-time job is enrolled on a one-day safety seminar and expected in that time to acquire all the knowledge necessary to keep his employers out of trouble. At least the appointee is sent on a training session, albeit that it only lasts one day!

Although there is growing recognition of the value of safety committees, firms who establish them frequently overlook the importance of suitable training for the committee members. This omission is remarkable given that the committee will be discussing matters which are heavily circumscribed by law and regulation.

In the worst case, this could mean that decisions are taken by the committee which cannot be justified in legal terms; more often time is spent discussing matters on which the committee is uninformed and therefore unable to make a valid contribution.

THE SAFETY COMMITTEE

Membership

The question of 'balance' is as important for the non-union company as it is for those with union representation, and it would be wrong to weight the membership in favour of management.

The guidance on safety policies (see Chapter 3) suggests that a director at board level should be assigned responsibility for promoting health and safety in the organization. It follows, therefore, that this director should be a member of the safety committee, and it makes good sense for him to act as chairman.

The attendance of the company safety officer is essential, since he will provide professional input and guidance. This resource can be used most effectively in two ways: first, to provide expertise, and secondly, to act as secretary to the committee. The importance of co-opting members from other fields of expertise has also been mentioned, but what of the 'lay' employee membership? How is it to be recruited? What qualities are needed? Will the members meet with the approval of the workforce? We will return to these points below.

Establishing the committee

Establishing a safety committee requires as much planning as other important business projects. Too many firms launch programmes which are colloquially referred to as 'the flavour of the month', a passing fad – and for many this could fairly describe their approach to the safety committee. It is introduced in a blaze of publicity and hype, yet within a few months has become a nonentity. Where passing enthusiasms are the norm, there is a significant attitude problem to be overcome – namely, the cynicism with which each new project is greeted.

A well-planned safety committee launch obviously requires effective written and verbal communication. Initially a message from the chief executive confirming the unequivocal support of the board, followed by publication of the programme for establishing the safety committee. As arrangements need to be made for the election of employee members, at least a month's notice should be allowed before the first planned meeting.

The director responsible for health and safety in the company should publish the follow-up communication, which should include the broad intentions for the safety committee, what the management representation will be, who will chair the committee and act as secretary, and so on. This announcement should include a timetable for election of members, and give the date, time and venue for the first meeting. The point that the company will arrange suitable training for members and that the training, as well as all committee meetings, will take place within normal working hours unless unusual circumstances arise, should be emphasized.

Employee membership

The number of non-management members will already have been determined. Ideally there will be sufficient to ensure that all parts of the business are represented, although this might not be possible in large multifunction enterprises.

The most democratic and popular method of recruiting employee committee members is to hold an election. Staff interested in serving should be asked to submit their names to the committee secretary to facilitate publication of CVs, photographs, and so on. It might become necessary for line management, with the assistance of the personnel department, to encourage employees to come forward either if it becomes clear that there may be insufficient candidates to cover all the vacancies, or to ensure that the election is contested. Election 'walk-overs' due to lack of employee interest are a bad start.

Some employees might be reluctant to offer their services through feelings of shyness or because they believe they have insufficient knowledge to make a useful contribution. Here again the point that adequate training will be provided for all the members of the committee must be stressed.

The willingness or otherwise of employees to serve on a safety committee can be affected by the state of industrial relations in a company. Yet a safety committee can be an opportunity to start afresh – to open a new chapter in labour relations. Much will depend upon the perceived sincerity of management, and not least upon the quality of the initial communication exercise. This more than anything else can make for success or failure.

Finally it makes good sense for the non-union company to appoint the elected employee committee members as the safety representatives for their areas. This enables them to raise concerns directly with the committee on which they have a voice.

The committee in action

The frequency, dates, time and duration of meetings over a six-month period should be decided at the first meeting of the committee, although it should be understood that a need for urgent unscheduled meetings might arise during that period. Frequency will be a function of the size and complexity of the enterprise. For some businesses, a monthly meeting will be needed, others may only require three or four meetings a year. Where subcommittees have been formed to consider particular issues, this might permit fewer meetings of the full committee.

The chairman should make clear – especially to the management members – that absence from meetings without real justification will not be accepted. The practice of managers delegating deputies to attend should not be countenanced. Nothing is more damaging to the effectiveness of a safety committee than the perception that some managers have decided it is not sufficiently important for them to attend regularly.

Meeting agendas should be produced at least one week before meetings, and minutes not later than one week after. Circulation of both should be to all members, with minutes also sent to the chief executive and members of the board.

Since January 1993, every employee has a statutory duty to report serious exposures or shortcomings in health and safety provision to his employer, or the employer's appointed safety advisers in accordance with MHSW (see Chapter 15). The best way to sustain interest in the safety committee – and therefore in occupational health and safety – is to keep the workforce informed about what is happening. This is not only good management, but encourages employees to be alert to bad safety practices, and to report them promptly. It might not be appropriate to post copies of safety committee minutes on notice-boards, but these should display a synopsis of matters discussed and actions determined.

Whichever course is chosen, it is essential that there is company-wide publicity for the safety committee.

Functions of the safety committee

The list of functions set out in Table 6.1 can be expanded to reflect particular circumstances, and may include any activity that relates to health, safety and welfare. The committee should not be asked to deal with matters which are unconnected with health and safety; to do otherwise could attract criticism from some sectors as well as diverting the committee from their *raison d'etre.*

Management response

The safety committee must have, and must be seen to have, a primary role in the organization's health and safety function. The acid test of management in this respect is their action – or reaction – to reports and recommendations from the committee. If there is a casual, indifferent or hostile reaction, or if senior management response is slow or negative, this will not only demoralize the safety committee, but word will spread rapidly through the business. The role of the committee chairman – who has a direct link to the company board – is vital in this respect.

Conclusion

Given the purpose in establishing safety committees, all the members, irrespective of their roles, must strive for objectivity. There is no room for 'them and us' attitudes, and if this is made absolutely clear at the outset, with subsequent events confirming it, the workforce will support the work of the committee, and believe in the sincerity of the company's health and safety message.

Table 6.1 Functions of the safety committee

- To study (and investigate and report where appropriate) accidents/incidents/statistics/trends.
- To review health and safety audit reports.
- To review reports of statutory inspectors.
- To develop works rules and safe systems of work.
- To oversee the health and safety content of training.
- To review publicity and awareness.
- To foster the health and safety message throughout the company.
- To provide a link with the enforcing authority.
- To consider proposed input to the safety policy update.
- To consider impending legislation and how it might affect the company.
- To ensure that their recommendations are acted upon.

Once a frank, open and positive climate is established, the committee can become involved with audits, reviews and the like. This will in itself improve morale, raise health and safety standards, and most importantly, knit together a group of keen and properly trained workers, whose support for the often hard-pressed cadre of professional safety staff will be of immense benefit to the business.

FURTHER READING

Health and Safety Executive (1976), *Safety Committees: Guidance to employers whose employees are not members of recognised independent trade unions*, no. HSC 8, London: HSE Books. Available free from HSE free leaflet line (telephone: 0742 892346).

HMSO (1972), *Safety and Health at Work*, the report of the Robens Committee 1970–72, London: HMSO.

Health and Safety Executive (1988), *Safety Representatives and Safety Committees*, (The Brown Book), London: HSE Books. Priced at £2.00.

MANAGEMENT ACTION CHECKLIST 6

Safety Representatives and Safety Committees

Checkpoints	Action required Yes	No	Action by
Is some action needed to ensure appropriate consultation with staff on health and safety matters?			
If a safety committee or equivalent exists:			
❖ Is there a proper management/staff balance?			
❖ Are there unfilled vacancies?			
❖ Does the committee meet regularly?			
❖ Are planned meetings ever postponed/cancelled?			
❖ Do management members delegate attendance to subordinates?			
❖ Are agendas circulated a week before meetings and minutes within a week following them?			
❖ What measures are in place to monitor the 'clear-up' rate of concerns raised by the committee?			
❖ Are company management regularly appraised of the work and effectiveness of the committee?			

7

CONTRACTORS

The biggest problems facing a company in respect to health and safety are not those posed by their own staff, but by the performance and actions of contractors working on their premises. Often the arrival of contractors on site, especially for contracts of a construction nature, could be likened to the arrival of the advance guard of the army of Attila the Hun. It might not be rape, but pillage certainly; like the marauding army, they are only passing through!

There are many reasons why this is so – and these will be discussed together with other issues most likely to cause employers (the clients) health and safety problems. This is followed by a review of the specific issues where a new building, or refit or alteration contract, is involved. Many of the recommendations made in this section could also be applied to other types of contract work. This chapter should be read in conjunction with Chapter 21 – The Construction (Design and Management) Regulations 1994 (CD&M).

CONTRACTOR RELATIONS AND HEALTH AND SAFETY

It has been clearly established by precedent case law that an employer (the client) cannot discharge his health and safety responsibility to a contractor simply by telling that contractor what the company's health and safety rules and arrangements are. The employer must devise ways to ensure that the contractor is complying with his safety policy, and that any subcontractors brought onto the site by the main contractor are properly trained and briefed, and that they also comply with the client's safety policy.

The only circumstance where the above principle does not apply is where the contractor is carrying out the contract in a separate, secured and delineated area, to which *only* his employees have automatic right of access and egress, and where the client's staff may only enter by appointment, and by 'signing in' as visitors.

Even then, access to and from the defined site by the contractor and those delivering goods to him must be over roads which have been assigned exclusively to the contractor for the duration of the contract, and are not shared with the client, his or her employees, or carriers delivering goods to the client. Where these conditions are satisfied the contractor takes complete responsibility for health and safety, including the necessary qualification and training of any subcontractors he employs.

WHAT IS A CONTRACTOR?

There are many answers and permutations of answers to this question. The following list is *only* a sample to illustrate the nature and variety of contractor activity in business today:

- Service engineer servicing or repairing his company's machines on a client's premises.
- Floral decoration contractor visiting to water or change plants.
- Taxi driver calling for a fare or dropping off.
- Building contractor erecting a new building on a client's site in a separate, delineated and secured area.
- Decorating contractor working around client's staff during normal working hours.
- Temporary 'holiday relief' staff.
- Staff supplied by an agency to work alongside client's own staff.

CONTRACTOR SAFETY PROBLEMS

Why do contractors and their operations create greater safety problems than one's own staff? Again, there is a variety of reasons:

1. They may not be properly or constantly supervised. For example, a small decorating contractor has a contract to paint offices by working around the client's own staff during normal working hours. On arrival on day 1 the boss tells them that he will be calling in at intervals during the week. However, the long-serving employees know from experience that this will not happen, and that they will receive the first visit ten minutes before finishing time on Friday afternoon for the purpose of paying their wages. They have been given the 'freedom' of the building, which means plenty of opportunity to chat to the client's staff, use facilities, and mingle generally.
2. Contractor's staff feel no loyalty towards the client; they are just 'passing through' – next month they are likely to be miles away.
3. Contractor's staff are probably not aware of the client's health and safety requirements. Although their management were thoroughly briefed, the information was not passed down to them.
4. The client has deadlines and wants the work done quickly, or the client alters the specification frequently during the course of the contract.

5. No liaison or communication between contractor's and client's staff/management.
6. No supervision or other check on the performance of the contractor by the client's own staff.
7. The contractor fails to brief his subcontractors properly.
8. The client's staff treat the contractor's employees as 'second-class citizens'.

ACCIDENT REPORTING

Every contractor should have an accident reporting procedure, and he should know about *all* the accidents that occur in his enterprise – wherever they occur. It is also the contractor's duty to comply with his statutory reporting duties in accordance with the Reporting of Injuries, Diseases and Dangerous Occurrences Regulations 1985 (RIDDOR) (see Chapter 9).

Notwithstanding, the employer (the client) needs to know about *every* accident and dangerous occurrence that occurs on his premises, irrespective of who is involved. An employer could hardly claim to be in control of his business if he did not know about these occurrences. If a contractor has a serious accident or reportable dangerous occurrence, this will precipitate an investigatory visit from the appropriate inspector. If the client is unaware of the occurrence the arrival of the inspector may prove embarrassing. As many accidents involving contractors' operations interact with the client's premises or equipment, it is important that the client has advance warning of the possibility of an inspector's visit.

CONTRACTOR HANDBOOKS/HANDOUTS

When a company uses contractors frequently, it is sensible to produce a handbook for contractors covering all the health and safety (and other) matters which the contractor should be aware of while working on the client's premises.

Some employers require every contractor's employee to receive a copy, which they sign for. This might be useful in the event of an accident or dangerous occurrence, if only to demonstrate that the employer (the client) had taken a responsible position with regard to contractor health and safety.

A shortened form of the handbook, perhaps comprising a pocket-sized card or booklet, could be used for occasional contractors (e.g. service engineers).

HOW THE CONTRACTOR PROPOSES TO DO THE WORK

While it would be wrong for a client to give his or her contractor precise instructions about how to do the work – especially if it required special skills/technology – it would be equally wrong to distance himself totally from what is going on. The contractor should be asked to provide a 'method statement' at the tender qualification stage, and checks should be made to ensure that in addition to compliance with the client's safety policy, the contractor is doing the job as he set out in the method statement.

INTERFACE OF CLIENT AND CONTRACTOR

Many companies channel all communication with their contractors through a designated employee, often styled the 'contract co-ordinator'. Where a number of contracts are running concurrently, there might be more than one contract co-ordinator. If all information, instruction and liaison is through the appointed co-ordinator(s), this should make for a smooth communication process, and avoid contractors receiving duplicate or conflicting instructions.

TEMPORARY STAFF

When temporary staff are required, these are usually recruited through temporary staff agencies to cover for holidays, staff absences or for work peaks. The temporary nature of the arrangements often results in the employing company overlooking or ignoring these staff where health and safety considerations are concerned. Often this is because the client does not want the temporaries to feel that they are in any sense part of the regular workforce since this might imply some 'permanent' status in accordance with employment law and practice. While this concern is understandable, it is extremely dangerous not to provide temporaries with a full health and safety briefing, and could expose the employer to legal action.

This matter is now specifically addressed by Regulation 13 of the Management of Health and Safety at Work Regulations 1992 (MHSW) (see Chapter 15).

USE OF TOOLS AND EQUIPMENT

If the work specification has been produced properly, the successful tenderer will have scheduled all the tools and materials needed to carry out the work. This being so, it should not be necessary for him to borrow anything. Immediately there is a request to borrow something, then, unless some totally unforeseeable situation has arisen, this should be seen as a warning signal – an indication that the contractor is less efficient than you thought.

In any event, the loan of tools to the contractor must be accompanied by more than a signature of receipt in the storeman's ledger. The borrower must also state clearly in writing that 'At the time of borrowing the following tool/equipment . . . I certify that it was in all respects safe, properly maintained and suitable for the purpose of . . ..' Of course, an indemnity of this kind does not provide a total defence in the event of things going wrong, but it will carry considerable weight.

Although not a safety matter, it would also be wise to ask the contractor to state in writing the circumstances in which borrowing the tool/equipment became necessary – or why he is unable to provide the item in question. This might be a matter for the client's purchasing department or others concerned with contract enforcement.

CONSTRUCTION WORK

This section covers the health and safety implications for a contract concerned with the construction of a new building or the refitting or upgrading of an existing building.

INTRODUCTION

The process

This section describes the more common arrangements when a new building is erected, or upgrade/refit takes place in an existing building. Much of what follows could also be applied to other contractor situations.

When significant work of a construction nature is planned, the usual practice is for the client's architect and/or mechanical or electrical consulting engineer to advise on the appointment of, or themselves appoint, the works contractors.

When the contract is completed, it falls to the resident maintenance staff, or an appointed specialist contractor, to maintain that which has been built, installed or modified.

Stages of the contract

For the sequence described above to work effectively and safely, it follows that those who will have to maintain the completed work should be involved or at least consulted during the stages of the contract, as follows:

Preliminaries During contractor selection their knowledge and experience may be valuable, perhaps to warn of a bad performance record.

During the tenure of the contract Continuous liaison to ensure that work in progress is compatible with ongoing operations of a routine nature (e.g. there may be a requirement to comply with a 'permit to work' system).

Handover Before the certificate of practical completion is signed, there should be a process of induction. The maintenance group should be satisfied that they are able to operate the new equipment and process with safety.

We will now examine these three stages or phases of the contract in some depth.

PRELIMINARIES

Contractor selection

In the days when safety was regarded as an adjunct to business, not an integral part of it, this paragraph would probably have called for considerations of safety to be assessed *after* the successful bid, based upon price, programme and performance, had been determined.

It must now be clear to all responsible employers that health and safety considerations are as important – in fact, more important – than any other element of the tender or tender qualification procedure. This view was upheld in a recent case, *General Building and Maintenance plc* v. *Greenwich London Borough Council*,* where General Building and Maintenance contested their disqualification from tendering for building maintenance work for Greenwich Borough Council. The plaintiffs held that it was wrong to exclude them during the first round of sifting the 'applications to tender' on the grounds that they did not meet the council's minimum health and safety requirements. Greenwich Council included health and safety matters as part of their evaluation of the 'technical capacity' of prospective tenderers. The court held that 'technical capacity' *did* include matters of health and safety, and therefore Greenwich Borough Council's refusal to allow the complainant to tender was upheld.

The key question, therefore, must be 'Does the tenderer have a responsible attitude and approach to health and safety?' Fortunately the client is able to ask a number of questions of prospective contractors to enable him to reach a decision on this point.

First and foremost is the submission of the tenderer's safety policy. If he employs five or more staff he has a statutory duty to publish a safety policy. If there is no policy, there is a breach of statute. Presumably an employer (the client) would not wish to enter into a relationship with such a company, particularly when the breach is germane to the question of health and safety in his premises.

There is another side to this question. Any sensible tenderer will also wish to have a copy of the client's safety policy. How otherwise can he be comfortable that his price takes account of all the particular health and safety requirements of the prospective client? There have been a number of 'claims' raised during construction contracts, where the contractor believed that he had the right to make a claim because his price did not include for the level of safety provision he subsequently realized was expected of him.

The safety policy is usually a good indicator of the overall standards of a company. Therefore its value extends beyond that of assessing the health and safety requirements of the tenderer.

To understand how much the client is involved in contracts where any work takes place within his existing site, we should consider the case of *R* v. *Swan Hunter Shipbuilders Ltd* [1981] IRLR 403. The case involved several fatalities which occurred when a subcontractor was working on HMS *Glasgow*, a warship undergoing refit at the yard of Swan Hunter.

In this case Swan Hunter, who produced adequate instructions on health and safety matters for their own employees, did not ensure that similar information was provided for their subcontractors, even though they had issued it to the managing contractor who they had retained for the refit contract. Among other matters, the safety instructions highlighted the dangers of oxygen-enrichment in the confined spaces of vessels, and warned of the importance of ensuring that all service supply lines were turned off at the end of shifts. Subsequently, the employees of one of the subcontractors engaged by the managing contractor caused an explosion and fire as a result of producing naked flames in an

*(1993) *The Times*, March 9, QBD.

oxygen-enriched atmosphere, and a number of workmen were killed.

In addition to the managing contractor and the employers of the deceased being prosecuted, Swan Hunter were also convicted in the Crown Court and their appeal to the Court of Appeal was dismissed. Essentially the Court of Appeal stated that if a danger to employees was recognized by their employer, and he issued warnings to them of that danger, it followed that the employer had the same duty to warn others who might be at risk. The appellants were prosecuted on indictment for a breach of Section 2(2)(a) and (c) and Section 3(1) of HASAWA (see Chapter 1).

Other relevant questions which an employer could put to a tenderer might include: a request for details of accident records over a stated period; numbers of enforcement orders served; details of organizational structure to manage health and safety; and, since January 1993, details of their appointed safety adviser (see Chapter 15).

Planning and briefing stage

Most construction work is carried out in one of two ways, as follows:

1. The contractor works in space occupied by the client, either by working around client's staff during normal working hours, or doing the work at evenings and weekends, or a combination of these.
2. The contractor takes over an area exclusively for his use during the contract. Client's staff may only visit by appointment, and will be treated in all respects as visitors, being required to sign an attendance book, and being accompanied at all times while on the site. Moreover, the site must be properly delineated and secured, and access to and from it must only be available to the contractor, his staff and those making deliveries to the site. If there is any shared use of the access road, this is *not* a separate site.

At the meeting held to co-ordinate the work once the contract has been awarded, the client should ensure the attendance not only of his technical professional advisers/consultants, but also of his own safety experts and representatives of the organization with ongoing responsibility for maintaining company buildings and plant. The reasons for safety representation need no justification here. Maintenance, who will have to maintain the buildings or services after the contractors have gone, must be able to contribute and to understand fully what is going to happen. The contractor should also have his own safety experts at this meeting. Further clarification of these roles and responsibilities will be available when the proposed new Construction (Design and Management) Regulations are promulgated (see final section of this chapter).

The contractor should produce a 'method statement' to describe how he intends to do the work. This will enable the client's team to highlight any practical problems (e.g. access, use of existing lifts, need for 'permits to work', and so on).

Compliance with the client's safety policy has already been mentioned, but two elements of this should be emphasized:

1. *Evacuation procedure* Contractors *must* comply with them, even if they are working outside, or on the external façade of the building. In the event of an emergency evacuation, the fire brigade will wish to know who is left inside the building when they arrive on the scene. If proper control over the movements or whereabouts of contractors is not exercised, this question cannot be answered.
2. *Accidents and dangerous occurrences* Although the onus for reporting these rests chiefly with the contractor (for his own staff), it is essential that the client is also advised of *all* accidents, near misses and dangerous occurrences which take place on his premises.

DURING THE TENURE OF THE CONTRACT

There must be regular meetings, especially those which the safety staff of both the client and contractor attend. The position of consultants and safety advisers will be further clarified by the forthcoming Construction (Design and Management) Regulations (see final section of this chapter). At these meetings, which must be minuted promptly, safety must be included on the agenda, and any person present should be able to raise health and safety matters, either on their own account, or on behalf of those they represent.

HANDOVER/COMPLETION

Before the final handover, the contractor must provide thorough briefings and training, including the provision of manuals as demanded by the contract, to the client's staff who will subsequently have to maintain the building or plant. Operating and maintenance manuals should be passed to the maintenance group in sufficient time before the departure of contractors to enable them to raise any points of concern or lack of understanding with them.

It is vital to the ongoing smooth and safe running of buildings and plant that these matters are *not* left until after contractors have left the premises. The cost of post-contract attendance by the contractor, and further training of resident maintenance staff, will be prohibitive compared to costs incurred if the briefing and training takes place within the contract term.

In addition to the technical completion of the contract, the client's maintenance and safety staff should be satisfied that it is appropriate to accept the work from an operational and safety standpoint. If they are not, they should advise their own senior management of the fact, who should make the decisions. The safety officer advises, the line decides.

CONCLUSION

Because of the nature of construction and allied work, it is likely that the contractor's employees will have more accidents than the indigenous staff. It is also true that contractors' employees may have less regard for the buildings in which they are working temporarily. For this reason, the employer's staff must be vigilant and report any excesses, misuse of fabric and facilities, and so on, that they note. Most importantly, they should report at once any dangerous practice or situation which they observe. This was already an implied duty in accordance with Section 7 of HASAWA, and is now reinforced positively by Regulation 12 of MHSW (see Chapter 15).

Furthermore, in the interest of good communication, as well as being a practical step to minimize danger, staff should be given proper warning of impending contractor activity. This should not be viewed as the premature disclosure of planned moves or company strategy, but as a matter of good safety practice. A responsible attitude towards contractor operations will not be achieved unless the client's own staff – and their consultants – understand that the company's position on health and safety is determined and uncompromising.

It is well established by case law that when problems develop in building and M&E-type contracts, indifference and expediency are often factors. Where this is true of contractors, it is frequently found to be ignored or even condoned by the client or his consultants. In such cases, both contractor and client have been accused of criminal liability, and subsequent civil lawsuits for compensation or damages have joined together the injured or deceased's employer *and* the client company in the action. In some of these cases, where responsibility is seen as being a joint one, plaintiffs might select the company best able to afford to pay as the target for their litigation.

Thus we have an area of business where contractors have a higher propensity for accidents than one's own staff, where many of the safety arrangements are 'looser' than elsewhere, and where an injured contractor's employee may select the client company as the target for their compensation claims. For these reasons the close and effective management of contract relations and safety arrangements is absolutely crucial.

EU-DRIVEN REGULATIONS RELATING TO CONSTRUCTION

Chapter 21 covers the Construction (Design and Management) Regulations 1994 (CD&M).

It is important to consider Chapters 7 and 21 together. While the CD&M Regulations now codify much of the philosophy of this chapter, which deals with contractors generically, it does not remove the requirement for a separate chapter on contractors. For example, a project which does not cause more than four people to be on site at any one time is *not* covered by CD&M; this means that many thousands of projects or tasks involving contractors will fall outside the CD&M net.

Even when CD&M does apply, this chapter on contractors will aid understanding of the reasons for the manner in which CD&M has been drafted.

Finally, this chapter has dealt with aids to contractor selection. CD&M requires clients to appoint competent Principal Contractors, and there could well be circumstances in which a client is asked to explain what checks were carried out to ensure the competence of a selected contractor.

Pages 64 and 65 discuss health and safety related questions which might be asked of prospective contractors. These questions are as valid for contractor selection in the context of CD&M as they are for contract work of a smaller nature.

MANAGEMENT ACTION CHECKLIST 7

Contractors

Checkpoints	Action required Yes	No	Action by
Is there a section on health and safety in the standard qualification and/or tendering specification of your company?			
Does this include questions relating to accident records, enforcement orders, and so on, and does it require tenderers to submit a copy of their safety policy?			
Who evaluates the health and safety responses?			
To what extent does the health and safety evaluation influence contractor selection?			
Is there a meeting of short-listed tenderers at which health and safety issues are covered?			
Do you publish a contractors' handbook setting out your rules and requirements?			
Do you have a policy regarding the loan of tools and so forth to contractors?			
Are your safety advisers and maintenance/premises staff involved in construction/refit contracts as follows:			
❖ in contractor selection;			
❖ in the preliminary meeting with the successful contractor;			
❖ during the tenure of the contract;			
❖ at the handover induction/testing/commissioning?			
Do you require that the contractor informs you of *all* accidents, dangerous occurrences and near misses that occur on your premises?			

PART TWO

BEFORE THE EUROPEAN UNION

❖

8

HEALTH AND SAFETY TRAINING

❖

Of all the health and safety responsibilities of employers, none is more neglected than training, in spite of the stress laid on training by government, the TUC, employers' organizations and trade federations. This could be due to the failure to recognize that safety training is as important – indeed, more important – than skills training, although all skills training contains elements of safety training.

Before considering the statutory training duty laid on employers, we should remember an important tenet concerning compliance with the law: namely, that ignorance (of the law) is not an excuse for breaking it. While this is an admirable concept which in other times might have been appropriate, today we must all be ignorant of some laws that affect us, simply because the sheer volume and complexity of those laws must defeat even the most assiduous. This is exemplified by the mass of EU-driven law of all kinds, but, in particular, by that relating to health and safety.

If all this law makes it difficult for employers to keep abreast, how much more difficult for the employee to do so. Whatever the level of safety staffing a company has – even if it is the most basic – that staff will be infinitely more able to understand and monitor changes in the law than untrained employees. Yet a great part of the new regulations are concerned with the duties of employees. And in addition to these new duties, employees also have some new rights. So if the employer is not going to inform and train his or her staff with regard to all these changes in law, who will?

Fortunately there is growing recognition of the need to inculcate the younger generation with the principles of workplace health and safety, and there is evidence that this is being done in some secondary and tertiary colleges and universities. However, this is by no means universal and can only be regarded as the tip of the iceberg!

In spite of the efforts of some enlightened employers it is clear that most employees are unaware of their health and safety responsibilities in law. Even today, 20 years after the ratification of HASAWA, few employees could clearly state what their statutory duties under this Act are, despite the fact that these duties are spelled out clearly and simply in the Act. If by now the provisions of HASAWA are not comprehended by most people in the workplace, what hope is there for the additional employee duties imposed by the Management of Health and Safety at Work Regulations 1992 (MHSW) (see Chapter 15), which have heaped complexity and confusion upon the simple and succinct English of Sections 7 and 8 of HASAWA, and which employees have to comply with in addition to their HASAWA responsibilities?

This is a sad state of affairs on two counts. First, because the employer who is not properly training his staff is in breach of his statutory duty, and secondly, because the opportunity to inform and instruct employees about their own duties and responsibilities is not taken, so that they remain largely ignorant on these matters.

Large numbers of employees – probably the majority – do not understand that they have a statutory duty to co-operate with their employer in all the measures which he takes to discharge his statutory health and safety duties and responsibilities, that they must look after themselves and others while at work, and so on. Of course, they understand that they might be breaking a company rule or procedure, but they would be surprised to learn that they were also breaking the law if they did not comply with their company safety policy or other instructions such as those for emergency evacuation.

Whatever else is missing in an employer's training arrangements for health and safety, he should at least make sure that everybody understands what their statutory responsibilities are, and what sanctions might be used against them if they fail in their duty.

THE LAW ON TRAINING

Although the employer's training duty is set out clearly in Section 2 of HASAWA, there has been much more reference to this duty – often in more precise terms – in the new EU-driven regulations. These references appear below.

Given the number and variety of training requirements, it will be inordinately difficult for an employer to justify failure to provide some health and safety training. Although our health and safety legislation is qualified by the term 'so far as is reasonably practicable', it is difficult to imagine how an employer could offer the excuse that it was not reasonably practicable to provide health and safety training, since there is little cost attaching to its provision, apart from that of time.

THE HEALTH AND SAFETY AT WORK ACT 1974 (HASAWA)

The responsibilities of employers to their employees appear in Section 2. Section 2(2)(c) calls for 'The provision of such information, instruction, training and supervision as is necessary to ensure, so far as is reasonably practicable, the health and safety at work of his employees'.

MANAGEMENT OF HEALTH AND SAFETY AT WORK REGULATIONS 1992 (MHSW)

Regulation 11 deals with capabilities and training. In summary it requires employers to take account of the health and safety capabilities of employees when entrusting tasks to them and to ensure that they are provided with adequate health and safety training, placing emphasis on induction, change of job responsibility, or the introduction of, or change in, work equipment. Regulation 11 also calls for repeat training, the adaptation of training to take account of changed risks to health and safety, and stresses that training must take place within normal working hours (see Chapter 15).

In addition to the general training described, employees should be familiar with all the six sets of EU-driven health and safety regulations that became effective on 1 January 1993. Therefore, all employers, however well they have trained their staff hitherto, must acquaint them with the contents of these new regulations. The following abstracts from some of the regulations deal more specifically with matters upon which employees must receive adequate instruction, information, training and supervision, according to HASAWA.

Management of Health and Safety at Work Regulations 1992 (MHSW)

Regulation 12 deals with employees' duties, which are additional to the duties in Sections 7 and 8 of HASAWA. The details of both sets of requirements are covered fully in Chapter 15.

Regulation 8 requires employers to provide employees with comprehensible and relevant information on the following:

- The risks to their health and safety identified by the risk assessment exercise demanded by Regulation 3 of MHSW.
- The preventative and protective measures adopted to combat the risks identified to comply with Regulation 3.
- The procedures developed to deal with serious and imminent danger as identified to comply with Regulation 7.
- The identity of those persons nominated by the employer to staff the procedures developed to deal with serious and imminent danger.
- Any risks which are notified to an employer by another employer regarding his operations where they share the same premises (Regulation 9(1)(c) refers).

Manual Handling Operations Regulations 1992 (MHO)

Regulation 5 calls for employees to make full and proper use of any system of work provided for their use by their employer (see Chapter 18).

Provision and Use of Work Equipment Regulations 1992 (PUWER)

Regulation 8 deals with information and training for staff, while Regulation 9 deals with the training of staff (see Chapter 20).

Health and Safety (Display Screen Equipment) Regulations 1992 (DSE)

There are numerous references to employee training and involvement in these regulations, in particular with regard to the thorough briefing of 'users' before they participate in the assessment exercise demanded by the regulations. One of the matters upon which users must be properly briefed is in respect to their entitlement to eye tests (see Chapter 16).

Personal Protective Equipment at Work Regulations 1992 (PPE)

Regulation 9 covers information, instruction and training of staff, while Regulation 11 deals with reporting loss or defect and requires employees issued with PPE to report immediately the loss of, or any defect in, the PPE issued to them (see Chapter 19).

INHIBITORS TO TRAINING

Before 1993, companies who did not provide any safety training for their employees would perhaps justify this for one or more of the following reasons:

1. *General safety training*
   (a) Perception that the workplace presented no hazards apart from the possibility of a need for emergency evacuation, which was practised once a year anyway.
   (b) Difficulty in identifying suitable trainers and planning the training.
   (c) Operational pressures prevented managers from planning and employees from attending training.
   (d) Belief that training did not produce a safer workplace.
2. *Specific safety training*
   (a) Expediency – job had to be done quickly.
   (b) Task/process was new. Working out how to do it safely was a process of trial and error!
   (c) Task appeared to present no hazards.
   (d) Persons carrying out the task/process viewed as experts on subject.

ACCIDENT AND DANGEROUS OCCURRENCE INVESTIGATION

Almost all injuries and dangerous occurrences reported to the enforcing authority in accordance with the RIDDOR regulations are the subject of investigation by them. Two questions are virtually certain to be asked at the beginning of the investigation: 'What was the system of work?' and 'What training did the injured person(s) receive to fit them for the work being done at the time of the accident?' It would be surprising and alarming if these questions were not put by the investigating inspector, yet the answers frequently reveal poor work systems and even poorer training, which of course incriminates the employer concerned.

TRAINING RECORDS

These records are indispensable not only to provide documentary evidence that training has taken place, but in order to ensure that training is not duplicated or overlooked. In short, to ensure that there is a structured approach to the training of employees.

THE PLANNING AND EXECUTION OF TRAINING

THE SAFETY POLICY

Some companies produce excellent safety policies that cover virtually all the key issues. These policies might include a thorough treatise on the health and safety duties of employees, which the company believes will suffice as 'training' and removes the need for training in the conventional sense. This could be true of many firms in the commercial sector.

Nothing could be further from the truth. Apart from the fact that documents of this kind are not always read, the real essence of the subject cannot be conveyed in a book. The safety policy is undoubtedly a most important part of the safety equation, and this fact *can* be conveyed to an audience during face-to-face training. But by itself, the policy is unlikely to prove an effective substitute for proper training.

THE TRAINERS

A line manager has responsibility for the health and safety of his subordinates, and it follows that he should train them. Where the work of the department is specialized, it might be that the supervisor or manager are the only persons competent to provide the necessary specialized training.

In a great many cases, however, the training does not have to be specialized, and line managers feel ill-equipped or ill at ease at the prospect of training their staff in health and safety. To be fair, it is unlikely that they will have been appointed for their lecturing ability, especially when the subject does not have instant appeal.

There are a number of ways to overcome this problem:

1. In any group of peer managers, there might be one who does feel comfortable about delivering health and safety training. The solution could be for this manager to provide training for a group of departments in exchange for some reciprocity in another sphere.
2. Where there is an employed safety manager/adviser, he could carry out the training. The problem with this approach is that the safety manager becomes overwhelmed by the numbers to be trained. Even where this solution is adopted, it should be mandatory that the line managers of those undergoing training are present, not only to confirm the importance attached to the subject, but so that they can familiarize

themselves with the subject, which is as much a matter for them as their subordinates.

3. A better use of the safety manager's time would be for him to develop training programmes and individual packages which could be utilized by line managers when carrying out the training. If each package comprised a lesson plan, with the subject broken down into logical segments, managers should be able to deliver the training effectively. The package might include 'master' foils, or be linked to selected videos.
4. Professional trainers/consultants could be retained. Although this can be expensive, it does enable larger groups to be covered, and ensures uniformity of delivery. A variation of this approach is to get professionals to train groups of one's own staff to carry the message forward – 'training the trainers'. External consultants could also produce training packages for use by line managers as described in item 3 above.

ENSURING THAT EVERYTHING IS COVERED

The importance of record keeping has been mentioned; no less important, given the amount and variety of training now demanded by the EU-driven regulations, is a training plan or programme to ensure that everything and everyone is covered. This plan will have to take account of operational requirements, shift and holiday patterns, sickness and other absences such as course or court attendance, and so on. So *careful* planning is most important.

Also, however strong the 'whip' to ensure that everyone directed to attend actually reaches the training venue, there will always be some who contrive to miss the session. These absences must be carefully recorded and, where appropriate, disciplinary action should be taken. There are always some who will try to persuade their supervisors or managers that they really are too busy to receive the training which the law demands they receive!

SUMMARY

There can be little doubt that health and safety training – however well presented hitherto – will require a complete reappraisal to ensure that it embraces all the additional requirements described in this chapter.

It can make a difference if employees understand that their role and involvement is central to the creation and maintenance of a safe and healthy working environment, and that by providing adequate training and information their employer is anxious for them to participate in the effort to improve health and safety standards.

MANAGEMENT ACTION CHECKLIST 8

Health and Safety Training

Checkpoints	Action required Yes	No	Action by
Does the company comply with its responsibilities for training in accordance with Section 2(2)(c) of HASAWA?			
If yes, are there documentary records to prove it?			
Who organizes health and safety training in the company?			
Who carries out the training – safety staff/line managers/other?			
Is there a formal written induction procedure?			
Does the procedure include a substantial health and safety element (e.g. for emergency evacuation, first aid, accident reporting, employee statutory health and safety responsibilities)?			
Are there regular checks on the induction programme, and do new starters sign to confirm receipt of the training?			
What are the arrangements for refresher/repeat training?			
What special health and safety training do managers at each level receive?			
What arrangements exist to upgrade the present health and safety training to cover the new/additional requirements imposed by new legislation?			
Training includes passing on information. Have all the DSE 'users' in the company been trained on the DSE regulations prior to co-operating in producing the DSE assessment forms, and do they know their rights *vis-à-vis* eye tests?			
Safety training is an important subject to which the company safety policy should make reference. Does it?			

9

ACCIDENTS AT WORK

No employer should question the importance of accident reporting, even if an accident results only in a minor injury – or no injury at all. Accidents, injuries, dangerous occurrences and 'near misses' are unplanned events – and every business ought to be concerned about *any* unplanned event that occurs on its premises or in connection with its operations. How else can it claim to be in control?

Accident intelligence ought to be as much a part of the information flow as the state of the order-book, cash flow, and all the other elements of business life. Why? Table 9.1 illustrates the potential of accidents to affect a business in an adverse and unwanted manner.

Morale

How can morale be maintained in the organization that is known to take little notice when its employees have accidents? One company, when confronted with the morale argument, said: 'We couldn't be accused of indifference, we give the telephone number of our insurers to any employee who is injured here.' Hardly a substitute for proper management attention and concern for employee welfare!

MOTIVATORS FOR ACCIDENT REPORTING

INDIRECT MOTIVATORS

In addition to the specific reporting requirements of the Reporting of Injuries, Diseases and Dangerous Occurrences Regulations 1985 (RIDDOR) – which are described in Appendix 1 to this chapter – there are elements of general health and safety duty which an employer would be unable to comply with if a properly developed accident reporting and recording system was not in place

Table 9.1 The potential effects of accidents on a business

Accidents *can* result in one or more of the following:

1. Loss of production.
2. Disruption to operational routines.
3. Additional administrative work.
4. Visit and investigation by the factory inspector or environmental health officer (EHO).
5. Unwelcome press and other media attention.
6. Loss of client/customer confidence.
7. Criminal prosecution and sanction.
8. Civil action for damages/compensation.
9. Other direct and indirect costs associated with:
   (a) loss or absence of expertise/experienced staff;
   (b) need to recruit temporary staff, pay double salaries;
   (c) reduced productivity/quality during learning curve; and
   (d) damage to product or raw materials due to inexperience and many other problems.

and being followed. These are in Section 2 of the Health and Safety at Work Act 1974 (HASAWA):

- *2(2)(a)* Providing and maintaining plant and systems of work which are, so far as is reasonably practicable, safe and without risks to health.
- *2(2)(d)* So far as is reasonably practicable, maintaining any place of work under his control in a condition that is safe and without risks to health, and so on.
- *2(2)(e)* The provision and maintenance of a working environment for his employees, that is, so far as is reasonably practicable, safe and without risks to health, and so on.

No employer could feel comfortable about compliance with any of the above if he was not aware of the accident record of the business, since this information would be one of the key indicators or measurements of compliance.

It is well established that there is considerable 'under-reporting' both of accidents in general and in some categories of reporting duty in particular. The HSE are taking an increasingly tougher view of such breaches, which not only have the potential to deny injured employees the full protection of the law in that inspectors would not have the opportunity to investigate the accident, but also deprive the nation of important accident and other data. By this it is not meant that the purpose of statutory reporting is to build up a national statistics file! Apart from the necessity for official investigation of serious accidents, the trends revealed by accident reports help to optimize the time of overstretched inspectors, by ensuring that guidance and resources are targeted properly.

Table 9.2 Some benefits of an effective accident reporting procedure

1. It is an aid to an employer in discharging his statutory duties in accordance with the Reporting of Injuries, Diseases and Dangerous Occurrences Regulations 1985 (RIDDOR). Failure to comply is a criminal offence.
2. It gives management timely information, and provides early warning of possibility of criminal or civil action against the company.
3. It demonstrates an employer's concern with regard to his common law duty of care.
4. It ensures that adequate records are kept to comply with statutory requirements.
5. It prompts an investigation with a view to ensuring that any necessary improvements to premises and/or procedures are carried out to minimize the risk of similar accidents.
6. It sustains employee morale. If accidents are overlooked or ignored, morale will suffer.
7. It means that the company is under control – management are in the 'driving seat'.

The information is also invaluable in indicating where new or improved guidance, or even regulations, are needed. This might be in relation to problems occurring across a wide spectrum, or in order to deal with specific work activities or processes, or to legislate for the special needs of particular groups of employees (e.g. young people, pregnant women). Table 9.2 indicates the benefits to be gained from an effective accident reporting procedure.

DIRECT MOTIVATORS

The Reporting of Injuries, Diseases and Dangerous Occurrences Regulations 1985 (RIDDOR)

Although there have been regulations calling for the reporting of some categories of injury for many years, the current regulations are more onerous than anything preceding them. They are not the easiest regulations to understand, and this creates problems in ensuring uniformity of compliance.

The role of the first-line manager

The natural organizational point for generating reports of accidents and so forth should be the first-level manager/supervisor, yet it is often the case that the first-level manager, far from being a 'key player' in the procedure, is almost incidental to it. Given that *every* manager has 'responsibility' for his subordinates, this is difficult to understand. No more compelling situation calling for the exercise of managerial skill and responsibility can occur at work than when an accident or injury occurs.

In many cases, the incident will occur in the department area, thus increasing the manager's need to become involved. Is the injured person receiving the best attention? Should their next-of-kin be informed? Was there misuse of equipment? Were the safe working practices being followed? These and other similar questions must be of concern to the effective and efficient manager.

To develop reporting arrangements which take no account of the first-level manager is not only dangerous legally, but also deprives such managers of the opportunity to develop and maintain skills which are key to their overall effectiveness. Considerable time and resource wastage can occur when those who manage at the 'sharp end' are ignored until events force their involvement. Sadly this may not happen until long after the accident, when the full benefit of their expertise and experience is reduced.

The argument for involving first-level managers at the outset is important for every accident, although there is a strong case to assign *external* reporting responsibility to one individual at each location. The external reporting requirements in accordance with RIDDOR are described in Appendix 1 to this chapter.

If every manager was charged with the *external* (i.e. RIDDOR) reporting responsibility in respect to those reporting to them, employers would need to provide detailed training on the regulations for all their managers. The logistics of this for the medium/large company would be extremely onerous – perhaps involving training for anything between 20 and 2,000 managers. Given the complexity of the reporting regulations, there would be a risk that sooner or later a manager would fail to make a required report, thus exposing the organization to prosecution. Moreover, provision of the training necessary for every manager to comply with the external reporting requirements would be a poor use of time, since many if not most managers will not experience an accident in this category during their entire working lives.

Thus a more effective solution, both in terms of best use of resources *and* in producing a uniform standard of reporting, is to appoint one person in each location to be responsible for the external reporting duty. The nominee(s) can then be trained uniformly, and if their identifies are known to all staff, the results should be altogether more reliable and efficient than if the duties were spread across the entire management spectrum. For the single-location organization, this means that only one person and deputy need be involved in external reporting.

RESPONSIBILITY FOR *EXTERNAL* (RIDDOR) REPORTING

Although no two organizations are identical, one of the following solutions is likely to suit most firms.

Safety officer or manager He or she will already have detailed knowledge of the statutory reporting requirements, and may already be maintaining other statistics (e.g. accident).

Office or administration manager Where there is no full-time safety officer, one or other of the above may have the duties of safety officer included in their job descriptions. Such office-holders must be 'competent' in accordance with the Management of Health and Safety at Work Regulations (MHSW) (see Chapter 15).

Personnel The responsibility may fall naturally into a part of the personnel function, who may, for internal administrative reasons, have to deal with the 'three-day incapacity' and 'disease reporting' parts of RIDDOR (see Appendix 1).

Accident reporting and recording cannot be effective without the commitment and co-operation of all staff. Managers and employers cannot report events about which they are ignorant, and it is a defence against a charge of failing to make a report which is demanded by RIDDOR that those responsible were unaware of the injury, disease or dangerous occurrence. However, there are two qualifications to this defence:

1. There must be a proper accident reporting procedure in the organization.
2. The procedure must be properly communicated to all staff.

Options for communicating the accident reporting system include induction and subsequent repeat training, including the procedure in the company health and safety policy, on notice-boards, and in the first-aid room, if there is one.

INTERNAL (COMPANY) REPORTING

THE DIFFERENCE BETWEEN *INTERNAL* AND *EXTERNAL* REPORTING

There have been frequent references to RIDDOR in this chapter, and further guidance on compliance with these regulations appears in Appendix 1 to this chapter. For ease of understanding, these regulations have been referred to as the *external* reporting requirements: that is, the employer has to report the events to an external authority.

Fortunately the great majority of accidents/incidents are not serious enough to warrant external reporting: that is, they qualify only for *internal* reporting. Various experts have attempted to quantify the numerical relationship between minor and serious accidents. The consensus of these theorists is that there are a substantial number of minor accidents for every major one: that is, in the order of 100 minor accidents to every major one that has to be reported 'externally'.

The difference between a minor and major accident may be simply a factor of time and place. The fitter's wrench falling from a high machine gantry onto the floor below may have little or no consequence if there is nobody walking beneath: but it will be quite a different matter if it falls onto the head of a passing worker. Yet the difference between the harmless and the serious event may only be a matter of a fraction of a second.

This is why *every* minor accident, near miss, and so on, is important, and should be the subject of a report – and possibly investigation. Unless the near misses and minor accidents receive appropriate attention, it can only be a matter of time before they become serious accidents, with possibly fatal consequences. It is well-established that to argue that an operation has been carried out many times without an accident is no defence against a charge of failing to take appropriate safety measures.

WHO SHOULD BE INCLUDED IN THE COMPANY (INTERNAL) REPORTING PROCEDURE?

It is *not* enough simply to keep details of accidents involving one's own employees. How can a company be in control when accidents are happening on their premises about which they know nothing? If your accident reporting procedure does not include the requirement to report *all of the following categories*, it is *not complete*, and you cannot claim to be in control of the business.

Accidents involving your employees

Comment: It is clearly important that employers know about *all* accidents and incidents involving their own staff, *wherever they occur in the course of their employment.* This includes accidents while travelling on company business, for example by car or public transport, while in hotels, on customer/prospect premises, and so on. Travelling to and from work is excluded, unless the journey is to a destination other than the employee's regular place of work, or the journey is by car for which the cost of fuel is reimbursed.

At an employee's normal work location, he is 'at work' once within the boundary of the worksite. This might be important in circumstances in which an accident occurs between the on-site car park and the office or factory, or on site roads.

All other accidents occurring on premises for which you have responsibility

Comment: You should want to know about *all* accidents on your premises, whoever the injured party, if any, is. Apart from one's own staff, this might include visitors (e.g. representatives, civic dignitaries, employees' families) *and contractors.* There are different external (RIDDOR) reporting requirements for contractors and visitors. If a contractor sustains an externally reportable injury on your premises, it is the duty of *his* employer to make the external report, whereas serious injuries to a visitor must be reported by the occupier of the premises in which the injury occurs.

Contractors may be employed because certain tasks are considered too hazardous for your own staff; this increases the propensity for accidents involving contractors, and emphasizes the importance of knowing about *all* the accidents that occur on your premises. The following examples illustrate this:

1. A catering contractor's staff may be prone to cutting themselves during food preparation, therefore increasing the risk of cross-infection, food poisoning, and so on. Knowing about this trend will not of itself prevent it, but the client company should be aware of the situation. They could bring pressure to bear for safer working methods or more care, no

doubt threatening that the contract might not be renewed if the number of accidents is not reduced. The point here is that it would be embarrassing to discover the cause *after* an outbreak of food poisoning!

2. If a contractor has an externally reportable accident on your premises, it is *his or her* duty to comply with the reporting duty. Normally this leads to a visit from the statutory authority to investigate. In many cases the contractor's activity on your premises will interfere or interact with your own staff and/or operations, and this will inevitably cause the investigating inspector to question your employees; early warning of this possibility is clearly desirable.

Accidents elsewhere than on your own premises

Comment: These accidents include those in which third parties have been involved/injured in circumstances which could give rise to a claim against you (e.g. third party injured by one of your employees in the course of their work).

An example might be where your staff are operating a stand on an upper floor of an exhibition hall, and part of the display falls onto the floor below and injures a member of the public. Although no injury has occurred to your own staff, and the accident is not on your premises, it makes sense for the accident to be reported to you, given the possibility of a claim for compensation and so on.

INSURANCE

Insurers expect employers whose risk they carry to act responsibly with regard to the health and safety of their staff, and to develop effective procedures to deal with statutory as well as internal accident/incident reporting.

Insurance premium size is not simply a factor of the number of claims made. Reduced premiums might be negotiated to reflect the positive approach taken by employers with regard to health and safety generally. Where this translates into effective measures to minimize accidents, promptly and accurately reporting when they do occur, and subsequently investigating to learn from the experience, there is a good case for negotiating better premiums.

Discussions with your insurers could bring about co-operation in producing a company accident report form that covers your requirements as well as those of the insurer. This in turn may result in a reduction in the amount of paperwork involved when there are accidents and subsequent claims. For example, a copy of every accident report form raised could be sent routinely to your insurer, who might accept this as the *only* form required in connection with any subsequent claim for compensation.

ACCIDENT REPORT FORMS

Do you need to develop an accident report form for your business? Currently you may require managers to report accidents by means of a note or internal

Table 9.3 Advantages in using a company accident report form

- The form itself acts as a motivator. It is one thing to omit to write a note or memo, another to fail to complete a prescribed form.
- The form concentrates the writer's mind. All the necessary questions are asked, and nothing is overlooked.
- Consistency.
- When an accident is reported, the questions on the form serve as a prompt, thus ensuring that all the key information is gathered at the outset.
- Reporting by means of a form facilitates computerized data collection.
- It saves time.

memo; this is not as effective or as beneficial as using a prescribed company form (see Table 9.3).

The benefit in producing a report form in collaboration with your insurer has been mentioned. If this course is followed, the form might contain a few extra questions, but this is compensated by using the form as the sole report of the accident, thus satisfying your requirements as well as the insurer's. If the injured employee subsequently seeks compensation, the original accident report remains the base document, with no requirement to make a further accident report to initiate a claim.

Appendix 2 to this chapter suggests the questions appropriate to a company accident report form.

RECORD KEEPING

Hitherto this chapter has been concerned with the reporting of accidents/injuries and so forth, and the introduction of company accident report forms. In medium/large businesses, however, more sophisticated accident information will be needed. The company accident report form can be designed to facilitate data entry, and there are many permutations of data that can be produced.

Discussion with your information systems (IS) manager will be helpful here, and the form introduced could be the product of a combined IS and safety team. Examples of information which could be gathered from each accident form include:

- age/sex of injured person;
- time of accident and shift (if applicable);
- whether or not machinery was involved;
- place where accident occurred;
- object causing injury;
- nature of accident (e.g. slip, trip, fall, struck by object, crushing, electric shock; and

- part of body affected;

in addition to a description of the accident itself.

A number of software houses offer accident data collection programmes. However, the creation of accident report forms, and programmes for producing accident and injury figures, are, of course, matters for consideration by each company. Every employer must maintain a record of externally reportable accidents and first-aid treatment administered. While many use the official book (BI 510), this is no longer sold by the HSE. In future it will be sufficient to maintain a record using a hardcover exercise book or similar. Details of the statutory first-aid treatment recording requirements appear in Appendix 1 to Chapter 12.

'NEAR MISSES' AND MINOR INJURIES

If management are to obtain meaningful or useful data from accident recording, it must be complete. As stated earlier, the difference between a serious, even fatal, accident and a 'near miss' might be a factor of time and space. 'Near misses' must therefore always be reported and, where appropriate, investigated.

There is a tendency to disregard minor injuries; for example, stubbing a finger on a drawing-pin. When accidents of this kind are ignored, it is certain that 'Murphy's law' will again apply. The plastered thumb or finger goes unreported, and in a few days the plaster is replaced by a finger cot. A week later, the arm is in a sling and hepatitis has set in – yet this was a minor accident deemed unworthy of reporting!

In addition to the above categories, some employers require accident forms to be completed where an accident, although causing no injury at all, has caused company assets to be damaged or destroyed. While the theme of this chapter has been concern for people, there is no reason why other assets should not be included in the reporting system. This not only makes for completeness, but accustoms staff to reporting everything out of the norm.

ACCIDENT INVESTIGATION

All accidents and incidents cannot be the subject of investigation, and every company will determine at what level of seriousness formal investigation will take place. There must of course be *some* investigation if things are to improve. It is pointless to keep detailed records if these are only used to pin-point particular areas of risk, or groups who are particularly accident-prone. While such information is important, it is only part of the equation.

Investigation can reveal where the systems and procedures necessary are inadequate or missing altogether. All too often it is only the immediate event which is the focus of attention, not the underlying cause or causes, and these are almost always to do with the failure of management or management systems.

Careful preparation is fundamental to accident/incident investigation.

Moreover, all those who will be interviewed must understand that the purpose of the investigation *is not* to apportion blame, but to consider ways to ensure that a similar accident does not occur. This is not always successful, and there are likely to be individuals who will withhold information from fear of incriminating themselves or others. Therefore it should be made clear – and should actually be the case – that the investigation is concerned with making improvements, and *not* with apportioning blame. As accident/incident reports are not privileged – that is, they may have to be revealed to opposing interested parties – it is essential that they do not in fact apportion blame in any way.

Unfortunately the whole approach to compensation in cases of this kind militates against health and safety, and some legal advisers have actually suggested refraining from making some improvements in case plaintiff's counsel uses the fact to prove that there were faults in the employer's premises or arrangements.

ACCIDENT/INCIDENT INVESTIGATIONS – SOME HINTS

- Start work as quickly as possible after the accident – memories fade, people collude.
- Apart from making the accident scene safe, try to leave it undisturbed. If this is not possible, or there is a chance of deliberate interference, take photographs.
- Obtain documentary evidence (e.g. test reports, certificates, maintenance records).
- Preface every interview with the assurance that the intention is not to apportion blame, but to establish the cause of the accident in order to prevent a recurrence.
- Interview witnesses separately, recognizing that in unionized concerns those interviewed may request the attendance of their union representatives.
- Present clear, concise reports.

FURTHER READING

Health and Safety Executive (1992), *Reporting under RIDDOR*, no. HSE24, London: HSE Books. Free from HSE Books, PO Box 1999, Sudbury, Suffolk, CO10 6FS (Telephone: 0787 881165. Fax: 0787 313995).

HSE/HMSO, *A Guide to the Reporting of Injuries, Diseases and Dangerous Occurrences Regulations 1985*, no. HS(R)23, London: HSE Books. ISBN 0 11 883858 X. Priced at £5.00.

HSE (nd), pads of Forms F2508 and F2508A, priced at £3.00 per pad of 10 forms.

APPENDIX 1

THE STATUTORY REPORTING DUTIES IMPOSED BY THE REPORTING OF INJURIES, DISEASES AND DANGEROUS OCCURRENCES REGULATIONS 1985 (RIDDOR)

IMMEDIATE NOTIFICATION – NORMALLY BY TELEPHONE

All fatal injuries and the following major injuries must be reported:

1. Fracture of the skull, spine, pelvis and any bone in the arm or leg; but not bones in the hand or foot.
2. Amputation of a hand or foot; or of a finger, thumb or toe where the bone or joint is completely severed.
3. Loss of sight in an eye or a penetrating injury, or a chemical or hot metal burn to an eye.
4. Injury requiring medical treatment or loss of consciousness due to electric shock.
5. Loss of consciousness due to lack of oxygen.
6. Decompression sickness.
7. Acute illness or loss of consciousness caused by absorption of any substance.
8. Acute illness believed to be the result of exposure to a pathogen or infected material.
9. *Any other injury which results in the person being admitted to hospital for more than 24 hours.*

WRITTEN NOTIFICATION

A written report of a fatal accident, or of any of the occurrences listed above, should be sent to the enforcing authority within seven days of any notifiable accident, *in addition to* immediate notification.

Also, there is a requirement to use HSE Form F2508 to report any other workplace injury to an employee *which results either in their absence from work for more than three days or their inability to perform their normal work for more than three days,* not counting the day of the accident, but counting every day thereafter, even if these days would not normally be working days.

The HSE are aware that there is considerable under-reporting of accidents generally, and of the three-day incapacity accident specifically. This is no doubt because employers tend to regard these accidents as minor. There are, of course, more of the three-day accidents each year than the sum of all the fatal and serious accidents, and HSE are taking an increasingly firmer line with offending companies.

DANGEROUS OCCURRENCES

There is also a requirement to telephone immediately – and follow up with written confirmation (again using HSE Form F2508) – in the event of any of the following dangerous occurrences:

1. The collapse, overturning or failure of a load-bearing part of a lift, hoist, crane, derrick or mobile platform, or an excavator, or a pile-driving frame with an operating height of over seven metres.
2. The collapse or failure of a load-bearing part of a passenger-carrying amusement device or any safety arrangements connected with it.
3. The explosion, collapse or bursting of any closed vessel.
4. Electrical short circuit or overload causing fire or explosion.
5. Any explosion or fire resulting in the suspension of normal work for more than 24 hours.
6. The sudden, uncontrolled release of one tonne or more of highly flammable liquid.
7. The collapse or partial collapse of any scaffold over five metres high.
8. Any unintended collapse of any building or structure under construction, alteration or demolition involving a fall of more than five tonnes of material or a wall or floor in a place of work.
9. An uncontrolled or accidental release or escape of any pathogen or substance from any apparatus or equipment.
10. Any unintentional ignition or detonation of explosives.
11. Failure of any freight container or a load-bearing part thereof.
12. Bursting, explosion or collapse of a pipeline.
13. Any incident in which a road tanker overturns or suffers serious damage, catches fire, or causes the release of dangerous substances.
14. Any incident in which a dangerous substance being conveyed by road is involved in a fire or where there is an uncontrolled release or escape of the dangerous substance.
15. Any incident where breathing apparatus malfunctions in such a way as to deprive the wearer of oxygen.
16. Any incident in which plant or equipment comes into contact with overhead power lines exceeding 200 volts.
17. Any case of accidental collision between a locomotive and train or any other vehicle at a factory or dock which might have led to death or reportable injury.

There are additional categories of dangerous occurrence in relation to mines, quarries and railways.

DISEASES

The disease reporting duty is a written one only, using HSE Form F2508A. However, it is for the patient's general practitioner to notify their employer that they have a notifiable disease, by endorsing the patient's medical certificate in red: 'Reportable disease'.

There are a number of reportable diseases which relate to specific work activities, and some which are 'general'. The general diseases covered by

RIDDOR are as follows:

1. Certain poisonings.
2. Some skin diseases such as skin cancer, chrome ulcer, oil folliculitis/acne.
3. Lung disease including occupational asthma, farmer's lung, pneumoconiosis, asbestosis, mesothelioma.
4. Infections including leptospirosis, hepatitis, tuberculosis, anthrax; and any illness caused by a pathogen.
5. Other conditions such as occupational cancer, cataracts, decompression sickness and vibration white finger.

If your business is commercial, retail, wholesale, hotel and catering, consumer/personal services, leisure and entertainment, or residential accommodation excluding nursing homes, telephone and written reporting should be to the environmental health department of your local authority. For all other premises, the report should be to the area office of the Health and Safety Executive (HSE).

In all cases the address of the statutory enforcing authority should appear on the 'Health and Safety Information for Employees Regulations' poster, available from HSE Books or Dillons bookshops, which should be prominently displayed in all workplaces. To ensure prompt reporting, this address should be carefully filed by the person responsible for reporting under RIDDOR in each company.

APPENDIX 2

ACCIDENT REPORT FORMS

Many companies have form design requirements, and accident report forms are capable of many permutations of layout.

There might be a preference for using the 'choices' approach to some questions: for example, 'Was the person involved in the accident permitted to be in the area in which it happened? – Yes/No'. This is the method used on the statutory reporting form HSE F2508.

This appendix therefore lists the important questions, which can then be arranged into any format, including data entry.

Details of injured person(s)

1. For employees: full name and title, occupation and personnel number.
2. For non-employees: full name, title, and name and address of employer (or own address if a visitor).

Details of the accident

1. Description of accident, including nature of injuries, if any; if eyes or limbs affected, state if left or right.
2. Date, time and place of accident.
3. Who was it reported by, and at what time/date?
4. Did injured persons:
   (a) cease work after the accident?
   (b) resume work after treatment?
   (c) lose time after the accident?

Witnesses

The names and addresses of all witnesses.

Hazard

1. If the accident was caused by an identified hazard, state what action has been taken to prevent a recurrence, and by whom?

Authorization

1. Was the injured person authorized to be at the place of the accident? Yes/No
2. Was he or she permitted to do the type of work on which engaged? Yes/No

DATA ENTRY

Where there is an accident record database, accident report forms will be designed so that much of the key information can be entered directly onto the database from the accident form itself. In such cases dates and times must be entered in a prescribed manner, and many questions have to be answered from a selection of options.

MANAGEMENT ACTION CHECKLIST 9

Accidents at Work

Checkpoints	Action required Yes	No	Action by

Does the company hold the following:

- Copy of the HSE definitive book on RIDDOR, *A guide to the RIDDOR*, no. HS(R)23 (see 'Further Reading')?
- A supply of reporting forms, no. HSE F2508 (see 'Further Reading')?
- An accident book at each location?

Is there a company accident form in use?

If there are more than 100 accidents of all kinds in a year, there ought to be an accident database record. Is there?

Is there a person nominated to ensure RIDDOR compliance at each location?

Has the nominee some HSE forms no. F2508 and details of the appropriate enforcing authority?

What is the method of communicating accident/incident reporting responsibility to all employees? Is it:

- induction training;
- included in the safety policy;
- refresher/repeat health and safety training;
- through first-aiders when they treat injuries?

Ideally all these methods should be employed.

Is there an accident/incident investigation process?

Are RIDDOR accidents/dangerous occurrences reported at board meetings?

10

CONTROL OF SUBSTANCES HAZARDOUS TO HEALTH REGULATIONS 1989 (COSHH)

The Control of Substances Hazardous to Health Regulations 1989 (COSHH) have been the subject of more intensive publicity than any other statute enacted since HASAWA became law in 1974. Despite this, there still remain companies in Britain who are unaware of COSHH, and many others that believe, quite wrongly, that they are not affected by it. Table 10.1 shows six degrees of response to COSHH. The reader might like to make an objective judgement as to which of the six categories his or her business falls into – recognizing that groups 4–6 must by definition be in breach of COSHH.

Undoubtedly there are very many firms and organizations that fall into the 'non-complying' groupings in Table 10.1, as HSE figures for COSHH enforcement orders and prosecutions confirm. Indeed, there is considerable concern about the extent of non-compliance with COSHH, despite the increased

Table 10.1 COSHH compliance – which category does your company fit into?

1. COSHH simply formalized your existing good practices.
2. COSHH was new, but your implementation was thorough. You feel comfortable.
3. COSHH was new, you intended complying and made a start, but somehow 'lost your way' or lost momentum. You do *not* feel comfortable.
4. You knew about COSHH, but have done nothing, believing that it does not affect or apply to your business.
5. You have heard about COSHH, but have done nothing, hoping that you won't be caught.
6. You have not heard of COSHH!

manpower granted to the HSE in order to monitor implementation.

As the requirement to produce a COSHH 'assessment' has existed since 1 January 1990, most businesses who do not have one are in breach of the regulations. There will be very few firms who have no need of an assessment of some kind. Therefore companies who should have carried out and published assessments, but have not done so, are at risk of prosecution, and the courts are taking an increasingly tough view of such failures.

As might be expected, official attention has moved from simply questioning the existence or otherwise of an assessment, to reviewing the actual assessments. Are they complete? Do they adequately address the hazards of substances used in the business? Are the systems and procedures described adequate to protect workers, and so on?

Why is it that the COSHH regulations have not been complied with more effectively and uniformly throughout the UK? It could reasonably be argued that failure has been due to businesses reaching 'saturation' point with the extent of health and safety regulation that has appeared in recent years. This coupled with the effects of recession could well have been the cause. It is much more likely, however, that the reason has been the unique character of COSHH – introducing as it does the notion that employers should themselves decide what precautions are necessary for their employees by taking account of the nature of the substances used in the enterprise, and the risks associated with them.

Thus after two centuries or so of law and regulation that is precise in requirement – and which leaves little or no room for interpretation – we have been confronted with a fundamental piece of health and safety legislation (indeed the most significant and far-reaching since 1974) that requires employers to make their own decisions about what is required. This basic shift in emphasis and responsibility not only 'wrong-footed' many, but revealed a vacuum in terms of knowledge, approach and experience. In these circumstances it would have been remarkable if firms and organizations had been totally and effectively responsive to COSHH from the outset.

Nevertheless the 'teething troubles' of the COSHH approach have their positive side. By conditioning companies to the notion of 'assessments', this regulation has provided a foretaste of what will be the usual approach to much of the health and safety, and indeed other legislation, of the future. Three of the EU health and safety directives which took effect from 1 January 1993 – the Framework Directive and those on display screen equipment and manual handling – include the requirement that assessments are carried out, and we may expect more of the same with other EU-driven legislation.

WHAT ARE THE OBJECTIVES OF COSHH?

There are many who view the COSHH Regulations as intrusive, time-consuming to implement, even unnecessary. Equally, there are many who feel uncomfortable in terms of compliance; they may have gone some way along the

COSHH path, but are not sure that they have gone far enough.

But it is more likely that most companies fall into the third category of Table 10.1. They know about COSHH, fully intended complying, made a good start, but somehow lost their way or their impetus. One company, when asked how they were progressing with COSHH, responded by saying 'Oh we've dealt with that, we have files containing data sheets for everything we use here – and they are in alphabetical order.' Further questioning elicited that nothing further had been done, the company concerned believing that keeping key information on file was all that was necessary! Indeed, a recent HSE report on COSHH compliance – *COSHH – the HSE's 1991/92 evaluation survey* – stated that:

> The most common fault was found to be that assessments consisted of little more than collections of data sheets, or other information about hazards on the premises, without an evaluation of the risks arising from these hazards.

There is some evidence for the view that for every company having a caring and responsible attitude with respect to the substances they use, there is another that is either ignorant or indifferent to the hazards to which their employees may be exposed when using them. Before and since COSHH there have been accidents, injuries, criminal prosecutions and civil claims arising from such ignorance or indifference.

The primary objectives of COSHH are therefore as follows:

1. A situation where worker exposure to hazardous substances is kept to a minimum commensurate with the needs of the business.
2. Employees to be told what residual dangers there are, and what measures have been taken to minimize exposure and danger.
3. Where some exposure to hazardous substances is unavoidable – and only after all reasonable measures to eliminate or reduce exposures have been taken – suitable protective clothing and equipment must be provided.
4. Where appropriate, employers must ensure medical review and collective feedback on the results for those employees exposed to hazardous substances.

RECORD KEEPING

In addition to the assessment itself, adequate records must be kept: for example, of the employees exposed to hazardous substances; of the information and training provided to such employees; the results of regular medical examinations (where required); and records of the inspection, testing, maintenance and effectiveness of engineering solutions adopted to minimize exposure (e.g. ventilation, improved exhausts). Finally, there should be a record of the issue and regular testing of personal protective equipment, including respiratory protective equipment.

MANAGING THE BUSINESS

Whatever the doubters may say, the plain truth is that COSHH is simply another element of the management of a business. Table 10.2, which summarizes the key elements of the COSHH process, confirms this.

KNOWING WHAT SUBSTANCES YOU HAVE

Many firms are unaware of the answer to this question. Often the COSHH inventory process deals solely with departments/functions who are the principal substance users, to the exclusion of other areas where the variety and quantity of substances is believed to be smaller, and by definition less important.

COSHH is concerned with *all* the hazardous substances present, however small the quantity. Where quantities are small, and hazards consequently reduced, this can be reflected in the assessment – for example, by not having such a sophisticated protection regime as for areas of greater risk.

Frequently the COSHH exercise reveals the existence of substances which, although hazardous, are no longer required. They may have been purchased for a process which was subsequently abandoned, and their existence forgotten, or, more likely, nobody has initiated action to dispose of them. This is the opportunity to get rid of unwanted substances, provided of course that their disposal is in accordance with local authority requirements and the Control of Pollution Act.

INVENTORY

Item 1 of Table 10.2 refers to 'substances' not 'hazardous substances'. This is deliberate. It is recommended that *before* the 'assessment' process is carried out, a complete inventory of *all* substances, whether deemed hazardous or not, is made.

Table 10.2 The key elements of COSHH*

1. Knowing what substances are used in the business.
2. Knowing what actual and possible hazards the substances used and processes followed create.
3. Reducing or eliminating the hazards.
4. Personal protective equipment (PPE) used as a last resort.
5. Information, instruction, training and supervision.
6. Medical surveillance and feedback (where necessary).
7. Regular maintenance and testing – engineering measures.
8. Maintaining the required records.

*Managing the business.

Table 10.3 Substance inventory

1. Obtain, read and evaluate manufacturers' data sheet for each substance/product.[1]
2. Record trade, generic and/or chemical name.[1]
3. User department/individual and extension number.[2]
4. Name, address and telephone number of manufacturer/supplier.[1]
5. Hazardous classification, if any.[1]
6. Exposure limits, if any.[1]
7. Purpose/use to which substance put.[2]
8. Mode of packaging and weight.[2]
9. Location of storage.[2]
10. Quantity held.[2]
11. Consumption rate.[2]
12. Manufacturer's recommended safety precautions.[1]
13. First-aid/medical treatment advised by manufacturer.[1]
14. Serial number of sheet (for administrative purposes).

Notes: [1]Information obtainable within the business.
[2]Information obtainable from manufacturers' data sheets, or dialogue with manufacturers.

A suggested format is shown in Table 10.3. Key to the process of completing the inventory are the data sheets produced by the manufacturer of the substance.

Every manufacturer (or importer/supplier) is required to supply users with a data sheet containing essential information about the product, highlighting any exposure standards (MEL/OES), the dangers, if any, inherent in its use, the protective measures that should be taken when using it, what to do in the event of spillage, and what first-aid/emergency measures should be taken in the event of inhalation, ingestion, skin contact, and so on. Data sheets expand on the basic information provided on the substance container.

Although there is a statutory duty to provide certain key information on the data sheet, the layout and presentation of the information is not at present stipulated. For ease of reference, and to facilitate production of computerized records, many firms opt to convert manufacturers' data sheets to a common format. This ensures that all the necessary information is available and presented uniformly to the reader.

Although the COSHH assessment must be carried out by 'competent persons', it is not necessary for those compiling the inventory to be 'competent' in the narrow sense, rather that they demonstrate commonsense and perseverance. If there is insufficient knowledge and experience available in the workforce, it might be necessary to employ consultants to carry out the assessment, which can prove expensive. The expense can be substantially reduced if the consultant assessor, instead of having to undertake a company-wide tour and compile an inventory, is provided with a detailed inventory of *all* the substances used in the business, presented in the format described in Table 10.3. With this information, the assessment process will be much easier, quicker and less expensive. This is not to suggest that an assessor can complete the work without getting about the premises in question, rather that much of the expensive preparation can be completed as a precursor to the assessment itself.

SAFETY POLICY

HSE guidance on company health and safety policies is that they should be periodically reviewed and updated. The guidance suggests that this should be done at intervals of about two years unless any of the following situations demand more frequent update:

- Health and safety organization within the firm changes.
- Processes change.
- Changes in law necessitate a review. In the case of the COSHH regulations, the guidance is that they are of sufficient importance to warrant a safety policy review/update.

The reasons for this are:

1. All employees must be aware of and understand the importance of the COSHH regulations.
2. All employees must understand that during the process of compiling the COSHH inventory and subsequent assessment, they must ensure that all substances which they are aware of (or have in personal desks/cupboards, and so on) are included in the inventory process.
3. That employees do not introduce or bring on to company premises *any* substance which is not approved in accordance with the COSHH review system determined by the company.
4. That employees do not use any classified substance which is not on the company COSHH assessment.
5. That every employee, irrespective of status, fully complies with the precautions determined in the company COSHH assessment for any classified substances which they use.

PERSONAL PROTECTIVE EQUIPMENT (PPE)

The emphasis of the COSHH regulations is towards the removal of dangerous substances (or the dangerous elements within a given product) in preference to simply issuing PPE. Indeed PPE should be regarded as a last resort, to be issued only when all other means of eliminating or reducing the hazard posed by the substance have been considered.

PPE must be suitable for the task and for the workers concerned. Those involved should fully understand the hazards of the substances they are using, and the measures which have been taken to minimize risks to their health and safety.

Employees must report immediately any damage to their PPE, or difficulty in using it. These responsibilities are reinforced in the new EU-driven PPE regulations (see Chapter 19).

The most common cause of injury/ill-health where PPE is concerned occurs, however, when it is not being worn at all! This problem is compounded where supervisors and managers do not deal firmly with staff who they observe not wearing the prescribed PPE. Many cases come before the courts where employees claim that their superiors were aware of their failure to wear the PPE prescribed, yet took no notice.

BEFORE THE ASSESSMENT

Table 10.4 is an *aide-mémoire* for those embarking upon an assessment for the first time, or reviewing the status of their existing COSHH assessment.

Item 3 in this table refers to operations and processes. It is important to remember that hazardous dusts are covered by COSHH, and these may arise not directly from a manufactured product, but from something which in its inert form is harmless. Woodworking is an example.

It is axiomatic that local exhaust ventilation (LEV) cannot be deemed effective unless this is proved by checking the air in the workplace. In all cases where air quality is suspect, some analysis must be carried out. This analysis will only be useful or valid if it can be compared to some baseline data. Therefore an air

Table 10.4 Before the assessment

1. Schedule all your substances, whether considered hazardous or not (the inventory).
2. Obtain manufacturers' data sheets for every substance on your inventory – manufacturers must provide them.
3. Consider what you do – your operations and processes.
4. Consider developing a company data sheet so that manufacturers' data sheets can be translated to a common format; facilitates computerization of data.
5. Consider your operations throughout the UK. Some economies in resource might be possible, as well as ensuring consistency.
6. Beware of locally held 'unauthorized' substances – however small the quantity.
7. Remember that contractors might bring substances onto your premises. They must furnish you with *their* assessments for whatever they store or use on your premises.
8. Review the safety policy. It must be updated to reflect your COSHH arrangements and procedures.

analysis regime should take the following factors into account:

- Are the products/processes in question subject to occupational exposure limits (OELs)?
- What were the measured conditions before LEV was installed?
- What are the conditions with LEV operating properly?
- Assuming LEV produces an acceptable standard, what arrangements exist to verify these are being maintained: that is, what ongoing analysis frequency is appropriate?
- What emergency analysis arrangements exist; for example, in the event of LEV malfunction/failure?

Occupational exposure limits (OELs)

There are two categories:

1. *Maximum exposure limits (MELs)* Substances having an MEL must be controlled so as to reduce them as far as is reasonably practicable, which must be to below the MEL.
2. *Occupational exposure standards (OESs)* Substances assigned an OES must be controlled so that the exposure is brought below the prescribed OES. However, control if the OES is exceeded will be deemed adequate provided that the employer:
   (a) has identified why the OES is being exceeded; and
   (b) is taking appropriate steps to comply with the OES as soon as reasonably practicable.

HSE Guidance Note EH40 lists OELs, and is updated annually.

FURTHER READING

HSE/HMSO (1988), *Control of substances hazardous to health (general ACOP) and control of carcinogenic substances (carcinogens ACOP): COSHH regulations*, 4th edn, London: HSE/HMSO. Priced at £4.50.

Health and Safety Executive (1993), *COSHH: A brief guide for employers,* leaflet no. IND(G)136(L), London: HSE Books. Free of charge.

MANAGEMENT ACTION CHECKLIST 10

Control of Substances Hazardous to Health Regulations (COSHH)

Checkpoints	Action required Yes	Action required No	Action by
By reference to Table 10.1, determine the degree of COSHH compliance.			
When was COSHH assessment last updated?			
Have any additional substances, or new or changed processes, been added since last assessment?			
Who is the custodian of the COSHH assessment?			
Is there a procedure to ensure that new substances are properly vetted in accordance with COSHH *before* being brought into the company?			
Does the safety policy make appropriate reference to COSHH?			
Are the following COSHH records kept as a minimum: ❖ assessment; ❖ data sheets for all substances; ❖ inspection/test of LEV, PPE; ❖ issues of PPE; ❖ COSHH training; ❖ air analysis (where applicable); ❖ medicals (where applicable)?			
Do contractors provide *their* COSHH assessments for *their* operations on your premises? (NB: As occupier you have overall responsibility.)			

Reproduced from *What Every Manager Needs to Know About Health and Safety* by Ron Akass, Gower, Aldershot, 1995.

11

FIRE AND FIRE PRECAUTIONS

Despite the fact that fire causes damage costing millions of pounds each year, and fire injury figures, although improving, remain unacceptably high, we find it difficult to take the subject seriously – it cannot happen to us!

The employer who is prepared to take risks may say 'We've been in business 30 years and we haven't had a fire', yet the truth is that for many firms, the fire that they have is the *only* one, with the result serious enough to put them out of business. Insurance might cover the physical losses, but it cannot replace experience or the records necessary to start up again. For a company with no records management programme or disaster recovery plan, the difficulties are daunting if not insuperable.

Despite the ever-present threat, the efforts of those responsible for fire prevention and precautions in a company are beset with problems – time, cost, indifference, apathy, getting the attention of executive management, and many others – and these are only the internal problems!

When contractors are working in the premises, these difficulties may be compounded. Among the unplanned situations that can arise – especially where electrical contractors are concerned – is that of fire alarms sounding due to inadvertent short-circuiting of the alarm system. Unfortunately, once this happens, it seems to continue day after day. Inevitably this leads to loss of confidence in the system, and building occupants become slower in their responses until eventually all sense of urgency has disappeared.

It is against this background that company fire officers and others with the responsibility for the management of fire emergencies have to operate. The lucky ones will have the support of the chief executive and his or her team, who ought to be giving a lead – setting the tone – for they have a greater responsibility for the health and safety of employees than anyone else. But all too often senior managers disregard fire practices – fire and smoke will only burn and choke lesser mortals!

LEGAL DUTIES – THE FIRE CERTIFICATE

A fire certificate is required for every business premises in which more than 20 people work, or where more than 10 work above or below ground level, although the issuing authority can grant exemption where they consider that the risk to occupants is insufficient to warrant certification.

Two other classes of business premises also require a fire certificate, even though they may not reach the occupancy thresholds mentioned above. These are:

1. Hotels and boarding-houses which provide sleeping accommodation for more than six people, or which provide sleeping accommodation above or below ground-floor level.
2. Where explosive or highly flammable materials are stored or used.

The certificate is usually (not always) issued by the Fire Authority, and it is an offence to occupy a business premises which meets the criteria described above without either possessing a fire certificate or having an official receipt for an application for one.

It is often the case that inspections for the purpose of considering the issue of a fire certificate can take place a long time – even years – after the application for one has been made by the building occupier. Subject to complying with the following requirements, premises may be used for the designated purpose while awaiting the fire officer's visit and inspection:

1. Ensuring that existing means of escape are maintained so that they can be used safely and effectively at all times.
2. Ensuring that all employees are trained in fire emergency procedures.
3. Ensuring that any existing fire-fighting equipment is maintained in efficient working order.

These conditions are not particularly onerous and a caring and responsible employer will want to ensure that his premises are brought to a standard equal to that demanded of a premises which has a fire certificate.

COMPONENTS OF THE FIRE CERTIFICATE

There are three main groupings:

1. Plans of the building which have been submitted by the applicant to the issuing authority, which the fire officer has endorsed to show which exits he deems to be fire exits, and where fire-fighting equipment and exit directional signs must be positioned. These plans must be kept with the certificate.
2. Any special consents that have been given, for example to store highly flammable materials. These consents will cover the position of the approved store, what fire-fighting equipment it must have, and the nature and quantity of the materials allowed to be stored therein.

3. Requirements for tests of fire-fighting equipment, fire alarms, fire training and evacuation, and so on, which are usually the same for all the business premises in the area. For each individual requirement there will be a stated frequency of test, and for each test or training session called for there must be an entry made in a 'fire log', which may either be a purpose-designed log issued by the county fire brigade, or where this is not the case an exercise book or ledger prepared and maintained by the certificate holder to record the tests, and so forth, carried out to comply with the fire certificate.

   Many companies neglect this important requirement, and, as fire officers' visits to premises are most infrequent, this omission may remain undetected. Needless to say, in the event of a fire and a subsequent insurance claim, the insurer would expect proof that the requirements of the fire certificate had been carried out. It is essential, therefore, that the fire log is kept up to date.

The fire certificate must be kept on the premises for which it was issued, and available for inspection by the fire officer.

It is an offence to carry out any material alteration to a building for which a fire certificate is issued, or to alter or vary any of the items for which specific approval was given – for example, in connection with highly flammable materials – without first obtaining prior approval to do so from the fire officer – preferably in writing!

The following are usually the subject of prescribed check or completion frequencies in the fire certificate:

1. *Training* Own staff; own staff and others who work in the premises outside normal hours (e.g. cleaners, security guards); persons not having English as their mother tongue; and illiterate persons.
2. *Equipment* Hosereels; emergency lighting; fire extinguishers; sprinkler systems; and fire alarms (usually weekly).

FIRE EMERGENCY PROCEDURES

The ultimate test of fire emergency procedures is the drill, the outcome of which must be a function of the quality of the management systems and arrangements made for emergencies. The following are the most important elements:

1. *Fire emergency notices* Are they simple, concise and coloured in accordance with current guidance, that is:
   (a) What *must* be done: white lettering on a *blue* background (e.g. what to do if discovering a fire or on hearing the fire alarm); and
   (b) What *must not* be done: white lettering on a *red* background (e.g. do not use lifts; do not stop to collect belongings; do not re-enter the building until instructed by the fire officer).

   These instructions are not exhaustive, and can be modified to suit local requirements. Surveys show, however, that fire emergency notices produced in this style, being clear and concise, and with colours corresponding to those designated in the Safety Signs Regulations 1980 (see Chapter 13), are much more likely to be read than other types.

2. *Calling the fire brigade* Are the arrangements understood by everyone?
3. *Arrangements for clearing the building* Are the arrangements for ensuring the building is clear of staff/occupants adequate? Roll-calls are unsuitable for large buildings because the fire brigade usually arrive before they are completed, and they want to know who is left inside. If this information cannot be to hand when the fire brigade arrive it is of little use. For large, heavily-populated buildings, it is better for floor wardens to account for the floors for which they have responsibility on a 'floor clear' basis, *not* by accounting for every person individually. This would put them at risk.
4. *Fire wardens/marshals* Are they up to establishment, trained and always present or seconded?
5. *Fire extinguishers* Is the company position clear regarding their use, and the training of those likely to use them up to date?
6. *Notices* Is there sufficient emergency signing, including directional signs, to ensure that staff and especially visitors to the building can get out quickly and safely?
7. *Exits* Do card-locked/security doors open automatically when the fire alarm sounds if they afford means to escape?

FIRE EMERGENCY DRILLS

Every employee – irrespective of their status in the business – must by law co-operate fully in all the measures which their employer takes to discharge his statutory health and safety responsibilities. Of the arrangements which an employer makes in this respect, none can be more important than those for fire emergencies. Emergency drills will be counterproductive, however, if they are organized without adequate planning.

The following factors relate to fire drill planning:

1. *Fire brigade* Do they wish to attend/supervise/be informed? Do you wish to invite them as advisers?
2. *Key staff* Who needs to know beforehand if they should remain in the building (e.g. business continuity in financial markets)?
3. *Weather* Is an alternative date earmarked in the event of really inclement weather?
4. *Exits* How to ensure all exits are used? What if some are shared with neighbours?
5. *Other factors* Assembly area discipline, especially if area is in a public place; banners for identification; members of public intermingling; problem of building re-entry, especially if high-rise; lift capacities.
6. *Communication* Arrangements for a 'wrap-up' meeting and

subsequent communication of details of the drill to staff/occupants.

THE TWO-STAGE ALARM

There was some resistance to two-stage alarms, but they have been introduced in many larger buildings to avoid the enormous amount of lost time occasioned by false alarms.

The arrangement is usually that the 'evacuation' signal only sounds in the area in which the alarm bell has been activated. Elsewhere an intermittent warning alarm sounds, alerting occupants that they may have to evacuate. In the standby areas staff can, if desired, prepare to evacuate by closing windows, and putting material into cupboards, while awaiting the evacuation signal.

Unfortunately, unless the building has a public address system, there is no way to announce that the emergency is over. Where there is no public address system the procedure is that staff on standby assume all is well after a determined interval without the evacuation signal.

SAFETY AND SECURITY

The safety of people is paramount. There have been cases where the concern of a company to protect its assets has had the effect of jeopardizing employees and others: for example, by having security arrangements/doors that do not 'fail safe' in an emergency.

It follows that in an emergency people must be able to use *any* method of escape that can be made available to them. It might be the case that a particular staircase becomes unusable, and escapers have to cross floors to another escape route. That journey must not be impeded by any locked door!

Examples of abuses and failure with respect to fire and fire precautions

- Complicated or poorly produced (even typed) fire emergency instructions. Sometimes out of date/obsolete or bearing the names of employees who are no longer with the company.
- Abuses/horseplay during emergency evacuation, including hiding to avoid evacuating – a criminal offence.
- Interfering with fire doors, usually by propping them open with the nearest fire extinguisher – a double felony!
- Obstructing escape routes – even momentarily.
- Making building alterations: (a) without obtaining the fire officer's approval (unless a minor change); and (b) without a review of the fire precautions arrangements that will be necessary (e.g. re-siting fire extinguishers and escape signs, checking the audibility of the fire alarm).
- New employee and temporary staff induction does not include full explanation of the fire emergency arrangements, nor a walk through all the escape routes available to them and a visit to their emergency evacuation assembly point.

- As above for contractors' staff.
- Rubbish stored in exit routes, often by the cleaning contractor. It remains there all day, putting the building occupants at risk, and is removed at night when they are no longer at risk. This is why a daily check of exit routes at the start of work is recommended.
- Interference with fire protection measures, including: extinguishers removed from hanging hooks/damaged/abused/wrongly discharged; also obstructed/unsigned/inaccessible/used to wedge fire doors open, and so on.
- Fire bells muffled with paper.
- Poor housekeeping. Paper left on desks and around office at night when it should be inside desks and filing cabinets. Lockable cabinets should be closed and locked at end of day, and failure to follow this routine could be enough to cause an insurer to repudiate a fire damage claim. Insurers expect that their policyholders will comply with their legal duty *and* established good practice.

Electrical equipment

This is now covered by the Electricity at Work Regulations (see Chapter 13). Fire certificates usually call for a complete inspection and test of electrical installations every eight years.

Smoking

Smoking in violation of a 'smoking prohibited' sign is clearly a breach of law. Those who smoke *and* those who observe smoking and do nothing are both in breach. Furthermore, research shows that a smoker is twice as likely to cause a fire as a non-smoker.

Identifying areas which should be designated 'smoking prohibited' is reasonably easy in industrial premises, but not so clear-cut in offices. Office areas with a high fire loading such as market trading floors, rooms containing large quantities of paper, and so on, ought to be designated 'smoking prohibited' on safety grounds.

In areas where smoking is permitted, special care is necessary. Cigarette ends and matches thrown into waste bins are a common cause of fire, as is the lit cigarette which is forgotten when work pressure increases (e.g. the smoker is called into a meeting).

CONCLUSION

The Management of Health and Safety at Work Regulations (MHSW) (see Chapter 15) require the development of procedures to deal with serious and imminent danger. In so far as fire is concerned, such procedures clearly exist already. MHSW is a reminder to re-visit fire emergency arrangements to ensure that they are up to date and provide the best possible system for preventing fire, and for getting people to safety if things do go wrong.

In many cases existing procedures will already have been honed to perfection, and will serve as a model for any additional procedures necessary to meet the requirements of MHSW.

FURTHER READING

Home Office/HMSO (1989), *Code of Practice for fire precautions in factories, offices, shops and railway premises not required to have a fire certificate,* London: Home Office/HMSO. Priced at £3.50.

Home Office/HMSO (1993), *Guide to fire precautions in existing places of work that require a fire certificate,* London: HMSO. Priced at £8.50.

Home Office/HMSO (1989), *Fire Precautions Act 1971: Fire safety at work,* London: HMSO. Priced at £4.00.

MANAGEMENT ACTION CHECKLIST 11

Fire and Fire Precautions

Checkpoints	Action required Yes	No	Action by
If your premises has a fire certificate could it be produced immediately on request?			
Who is responsible for the fire log, and for fire precautions in the company?			
Is the fire log up to date and showing that tests, checks, and so on, are being carried out at frequencies called for in the fire certificate?			
Has a fire drill been held within the past 12 months?			
If employees were randomly questioned, how many would know what they should do in the event of (a) discovering a fire, and (b) hearing the fire alarm?			
If fire wardens or marshals are part of your emergency evacuation plan, are they (a) recruited to establishment, and (b) do they have deputies to cover when they are absent?			
Are fire marshals and deputies formally trained?			
Does your induction procedure include a thorough 'fire' briefing, including a walk through all the available exits, *and* a visit to the assembly point for the inductee(s)?			
Do your daily inspection arrangements include a check to ensure that all fire routes and exits are clear of *any* obstruction, however small?			
When you conduct a fire drill, do *all* occupants leave the building apart from nominated key staff required to remain to attend to essential services?			
Does your evacuation plan include arrangements to hold a post-evacuation 'wrap-up' meeting and a communication to staff on the success or otherwise of the drill?			

Reproduced from *What Every Manager Needs to Know About Health and Safety* by Ron Akass, Gower, Aldershot, 1995.

Checkpoints	Action required Yes	No	Action by
Whenever changes in layout are made, is the effect upon existing exit signs, extinguishers, audibility of alarms, and so on, taken into account *at the planning stage, not after staff are 'in situ'?*			
Do the plans attached to your fire certificate reflect the present layout: that is, have you made any significant alterations that have not been approved by the fire officer or district surveyor?			
Are there any people working in 'dead ends'? Does everyone have two routes to escape in different directions?			
Are two routes to escape clearly visible, that is, by adequate signs that can be clearly seen from any point in a corridor or factory aisle?			

12

FIRST AID

Although there is growing recognition of the importance of proper first-aid provision at work, many employers have difficulty in deciding what that provision should be. This is hardly surprising. In earlier times it was simple enough. A table existed showing a ratio of first-aiders to employees, with some allowance for offices compared to factories. 'X' employees equalled 'Y' first-aiders and that was that.

In these enlightened times, employers are meant to be able to make such decisions for themselves – with guidance, of course. The latest guidance (1990) tells us that a calculation based purely upon numbers employed is not the best approach; there are other factors which should be taken into account.

PROVIDING FIRST AID

Before considering these factors, it is useful to examine an employer's statutory duty in this matter; this is contained in Regulation 3(1) of the Health and Safety (First-Aid) Regulations 1981, which states: 'An employer shall provide, or ensure there are provided, such equipment and facilities *as are adequate and appropriate* in the circumstances for enabling first-aid to be rendered to his employees if they are injured or become ill at work' [emphasis added]. It is as *simple* as that – or is it?

Having said that employee numbers alone are not a good basis for determining what is adequate and appropriate, the guidance then goes on to state that *every* employer must provide an 'appointed person' as a minimum *and* not less than one first-aider for every 50 employees. The guidance goes on to say that this ratio is thought sufficient for low-risk workplaces such as offices and libraries – the implication being that a much higher ratio of first-aiders ought to be provided where the risks are higher. In case this is too simple, the final word is that in hazardous workplaces nothing less than one first-aider for every 50 employees will suffice. So much for getting away from numbers!

The other factors which should be considered when determining what is 'adequate and appropriate' are the distribution of employees within the establishment, the nature of the work, the size and location of the establishment, whether there is shift working and, finally, proximity to professional medical services. It cannot be denied that all these factors, in addition to the number of employees, are germane, but without any kind of comparative reference, how do these considerations help an employer come to a *meaningful* decision – unless the employer is himself an experienced first-aider or is medically qualified!

Fortunately there is an agency which is responsible for providing advice and guidance to businesses on first aid, and indeed all occupational health matters – the Employment Medical Advisory Service (EMAS), which is part of the Health and Safety Executive. The address of the EMAS office responsible for the area in which the reader's business is located can be found on the statutory 'Health and Safety Information for Employees Regulations' poster which should be prominently displayed in every workplace. The address of EMAS appears in the box below that of the statutory enforcing authority responsible for health and safety in the premises.

COMMUNICATION

Having decided what first-aid provision is appropriate, the employer must: (a) provide it; and (b) tell his employees what the arrangements are, and how to obtain first aid. This communication could be on notice-boards, in the house telephone directory and safety policy, and, in the case of new employees, must form part of the induction process.

AVAILABILITY

It is important to ensure that the provision determined is available at all times that work is taking place. For example, if there is no shift working, and the agreed provision is two first-aiders, then two first-aiders must be present at all times. Only emergency situations can justify reduction in the level of cover, and emergencies *do not* include holidays.

This means that when planning first-aid provision, the following must be taken into account:

Holidays

How will the level of cover be maintained when a first-aider is on holiday?

If the business has a complete 'shut-down' for holidays, this is not a problem, unless there is a practice of bringing the maintenance staff in to overhaul plant and so on, in which case adequate first-aid provision must be made for them.

In all other cases the options will be to train and qualify more first-aiders than are necessary to cover holidays, or to employ a 'locum'. The former course has the merit that it creates a pool from which to replace vacancies caused by resignations, leavers, and so on.

Whether or not a pool or reserve of first-aiders is established, it is important to maintain a list of volunteers who are prepared to train, so that these can be fed into the training schedule in time to replace leavers. This is clearly a 'reasonable' measure which an employer can take to ensure that the cadre of first-aiders does not fall below the number which he has decided are necessary.

Emergency absences

In the event of absence due to emergency such as sickness, accident, and other reasons, the 'appointed person', whose role is described later in this chapter, is deemed to satisfy the employer's duty in terms of adequate first-aid provision.

Shift working

Where there is shift working, those on shifts must have the appropriate level of first-aid cover. Often, in particular for night shifts, this is difficult for the employer to arrange. In these circumstances it might be possible to utilize the services of security guards if they are trained and certificated. Many security companies provide guards who are qualified first-aiders as part of their service.

Exemptions

A company that has a resident doctor and/or qualified occupational health nurses may claim exemption from compliance with the requirements of these regulations.

First aid for persons who are not employees

There is no *duty* placed upon employers to provide first aid to persons other than their own employees.

However, it would be unusual not to make the first-aid provision available to *any* person who needed it while on the premises, and there is nothing in these regulations which prevents this. In the case of cinemas, theatres, shops, etc., it would be callous in the extreme, apart from commercially naive, not to make any first-aid facilities that exist available for patrons or customers.

Other options for first-aid provision

Where there is more than one employer on a site, or in a building, there is nothing to prevent co-operation in respect to first-aid arrangements. A simple written agreement should be drawn up setting out the arrangements, and each employer should possess a copy. It is the responsibility of each party to the agreement to inform their employees of the arrangements made.

In some places, for example industrial estates, there is often a central surgery, to which user companies subscribe if they wish. Some of these facilities are very sophisticated, and offer a wide range of health services in addition to first aid, such as servicing the first-aid boxes of subscriber companies.

FIRST-AIDERS AND APPOINTED PERSONS

The contribution of first-aiders and appointed persons is frequently overlooked by management and colleagues alike. They are volunteers, yet they have to train, pass an examination, and, once qualified, are virtually on 'standby' throughout the working day. The lucky ones receive a small honorarium, or a recognition lunch, or the like, but for many it really is a labour of love.

Volunteers should be temperamentally suited to the work, capable of undergoing four days of intensive theoretical and practical training followed by examination. As part of the first-aid provision of the company, they must be able to drop their normal work immediately they are called upon – not a job for the air traffic controller!

The first-aid course, including examinations, should last four days. Courses are conducted by organizations which have to be approved for the purpose, and include the national organizations such as St John's Ambulance Brigade. The qualification is valid for three years, before the expiry of which time first-aiders must take a refresher course of two days including re-examination. If this is not done before the expiry of the current certificate, the complete four-day course must be undertaken and the examination passed again. First-aiders may not provide first-aid treatment if they are 'out of time'. Ideally all first-aiders should receive refresher training each year to keep them up to date with the latest thinking and techniques. The principles of first aid have changed dramatically in recent years, and the process of change is likely to continue.

No treatment other than first aid for which they are trained may be given by a first-aider, and no medicines such as analgesics may ever be prescribed by them. Neither may medicines be stored in first-aid boxes, which must only be stocked in accordance with the guidance in the Approved Code of Practice for First Aid (ACOP). If a business has particular hazards for which the recognized first-aid training course does not provide training, it is the employer's duty to arrange special training, which can usually be provided by the agencies approved by HSE for the purpose.

Appointed persons

Where it is not considered necessary to provide a trained first-aider because the premises are small and the risks minimal, or where in exceptional circumstances the trained first-aider(s) may not be present (the circumstances *must* be exceptional – planned holidays are not included), the employer may nominate an 'appointed person(s)' whose duties are simply to provide emergency first aid and summon professional medical assistance, or arrange evacuation to a medical facility. Even if, in the employer's opinion, there is no requirement to provide any first-aiders, he must none the less nominate an 'appointed person'.

Emergency first-aid treatment may only be rendered by an 'appointed person' if that person has been trained to give it. There are short one-day courses specifically designed for 'appointed persons', which should be repeated every three years as a minimum.

FIRST-AID FACILITIES

FIRST-AID ROOMS

As with the number of first-aiders, the provision or otherwise of a first-aid room is no longer simply related to the numbers employed. Formerly it was necessary to provide a first-aid room if there were 400 or more employees. Now the criteria are similar to that for establishing how many first-aiders are needed – basically it is the employer's decision.

Many employers find it easier *not* to provide a first-aid room, but rather to identify a 'rest room' that may then double for other purposes, which a designated first-aid room may not. If a first-aid room is provided, however, it must be sited and be equipped to the standards in the Approved Code of Practice for First Aid (ACOP), which are described later in this chapter.

Regular cleaning of the first-aid or rest room is important. Frequently, due to its relatively low or non-routine usage, it is omitted from the cleaning schedule, with the result that this room – which should be cleaner than any other – is often the one that is rarely, and sometimes never, cleaned!

RECORD KEEPING

It is a specific requirement of the first-aid regulations – and an important safeguard – to maintain suitable records of first-aid treatment given.

Appendix 1 to this chapter shows the information required to be kept. It is important to make an entry in the record book of *all* treatment given *and* of any case where the injured person either declines treatment altogether, or opts not to receive treatment from the first-aider, preferring to go to his or her GP or a hospital instead.

FIRST-AID BOXES AND KITS

These boxes and kits should contain only a sufficient quantity suitable first-aid materials and *nothing else*. The contents should be replenished as soon as possible after use. Items should not be used after the expiry date shown on the packets.

There is widespread misunderstanding about the absence from first-aid boxes of tablets, medicines and other items for the treatment of minor illnesses. This is because the treatment of minor illnesses does not form part of the training of first-aiders, and they cannot, therefore, treat them. For this reason the HSE guidance notes on first aid make no reference to equipment for, or treatment of, minor illnesses.

Contents of first-aid box

- One guidance card as shown in Appendix 2 to this chapter.
- Twenty individually wrapped sterile adhesive dressings (assorted sizes).
- Two sterile eye pads, with attachment.

- Six individually wrapped triangular bandages.
- Six safety pins.
- Six medium-sized individually wrapped sterile unmedicated wound dressings (approx. 10 cm × 8 cm).
- Two large sterile individually wrapped unmedicated wound dressings (approx. 13 cm × 9 cm).
- Three extra-large sterile individually wrapped unmedicated wound dressings (approx. 28 cm × 17.5 cm).

If mains tap water is not readily available for eye irrigation, sterile water or sterile normal saline (0.9 per cent) in sealed disposable containers should be provided. Each container should hold at least 300 ml and should not be re-used once the sterile seal is broken. At least 900 ml should be provided. Eye baths/eye cups/refillable containers should *not* be used for eye irrigation.

Where there are special hazards for which one or more first-aiders in the company have received training in first-aid treatment, the special antidotes or special equipment for that treatment may either be stored near the hazard, or kept in the first-aid boxes.

Travelling first-aid kits

Where employees have to travel away from their home base extensively, and are likely to be in areas where medical assistance is some distance away, they should be issued with travelling first-aid kits. These kits should contain the following as a minimum:

- Card giving general first-aid guidance given (see Appendix 2 to this chapter).
- Six individually wrapped sterile first-aid dressings.
- One large sterile unmedicated dressing.
- Two triangular bandages.
- Two safety pins.
- Individually wrapped moist cleaning wipes.

SUPPLEMENTARY EQUIPMENT

Where an establishment covers a wide area or is divided into a number of separate self-contained units or working areas, it may be necessary to provide suitable carrying equipment for the transportation of casualties.

Where blankets are provided, they should be stored alongside the equipment and in such a way as to keep them free from dust and damp.

Disposable plastic gloves and aprons and suitable protective equipment should be provided near the first-aid materials and should be properly stored and checked regularly to ensure that they remain in good condition. Blunt-ended stainless steel scissors (minimum length 12.70 cm) for cutting away clothing should be kept together with the protective clothing and equipment, *not* in the first-aid box.

Plastic disposable bags should be available for soiled and used dressings.

These should be disposed of in sealed bags in accordance with the instructions of the local authority.

SITING AND EQUIPPING THE FIRST-AID ROOM

If an employer decides that a first-aid room should be provided as part of his or her overall arrangements, the room should be sited, prepared and equipped in accordance with the Approved Code of Practice (ACOP), as follows:

1. A suitable person (a trained and certified first-aider, medical practitioner or occupational health nurse) should be responsible for the room and its contents, and should be available at all times that employees are at work.
2. The room should be readily available at all times that employees are at work: that is, it must not be used for any other purpose than first aid or medical screening.
3. The room should be sited as close as possible to exits to enable the evacuation of patients to hospital. It should be large enough to hold a couch, with space for people to work around it, and a chair.
4. The room's entrance should be wide enough to accommodate a stretcher, wheelchair or carrying chair.
5. The room should contain suitable facilities and equipment, have an impervious floor covering and should be effectively ventilated, heated, lit and maintained.
6. All surfaces should be easy to clean. The room should be cleaned each working day and suitable arrangements made for refuse disposal.
7. There should be a suitable waiting area provided with chairs.
8. The room should be clearly identified as the first-aid room by using signs as specified in the Safety Signs Regulations 1980, and a notice should be attached to the door of the room showing the names, location and telephone numbers of the nearest specified/appointed person.
9. Other considerations are proximity to toilets and the possible need for emergency lighting.

Facilities and equipment

Facilities and equipment provided in the first-aid room should be as follows:

- Sink with running hot and cold water always available.
- Drinking water, when not available on tap, and disposable cups.
- Soap.
- Paper towels.
- Smooth-topped working surface.
- A suitable store for first-aid materials.
- First-aid equipment equivalent in range and standard to that shown for first-aid boxes (q.v.).
- Suitable refuse container lined with a disposable plastic bag.

- A couch (with a waterproof surface) and frequently cleaned pillow and blankets.
- Clean protective garments for use by first-aiders.
- A chair.
- An appropriate record book (see Appendix 1 to this chapter).
- A bowl.

Where special first-aid equipment is needed, this may also be stored in the first-aid room.

This lengthy schedule of requirements is one of the reasons why employers opt for a 'rest room' where they are uncertain. There must, nevertheless, be a first-aid room in large industrial companies, especially where doctors and/or occupational health nurses are employed.

FURTHER READING

HMSO (1990), *HSE Approved Code of Practice (ACOP): Health and Safety (First-Aid) Regulations 1981*, rev. edn, London: HMSO.

Health and Safety Executive (1990), *Revised 1990 First-aid Needs in your Workplace: Your questions answered*, rev. edn, no. IND(G)3(L), London: HSE Books. Available from HSE Information Centre, Broad Lane, Sheffield S3 7HQ (Telephone: 0742 892345).

APPENDIX 1

FORMAT FOR RECORDING FIRST-AID TREATMENT

Full name & address of persons who suffered an accident (1)	Occupation (2)	Date when entry made (3)	Date and time of accident (4)

Reproduced from *Essential Health and Safety for Managers* by Ron Akass, Gower, Aldershot, 1994.

Place & circumstances of accident – (state clearly the work processes being performed at the time of the accident) (5)	Details of injury suffered and treatment given (6)	Signature of person making this entry (state address if different from column 1) (7)

Reproduced from *Essential Health and Safety for Managers* by Ron Akass, Gower, Aldershot, 1994.

APPENDIX 2

GUIDANCE FOR FIRST-AIDERS

HEALTH AND SAFETY (FIRST-AID) REGULATIONS 1981

General guidance for first aid at work

NOTE: TAKE CARE NOT TO BECOME A CASUALTY YOURSELF WHILE ADMINISTERING FIRST AID. USE PROTECTIVE CLOTHING AND EQUIPMENT WHERE NECESSARY.

TREATMENT POSITION

Casualties should be seated or lying down when being treated, as appropriate.

Advice on treatment

If you need help send for it immediately. If an ambulance is needed, arrangements should be made for it to be directed to the scene without delay.

Priorities in first aid

(1) **BREATHING**

IF CASUALTY IS NOT BREATHING

Place on back.

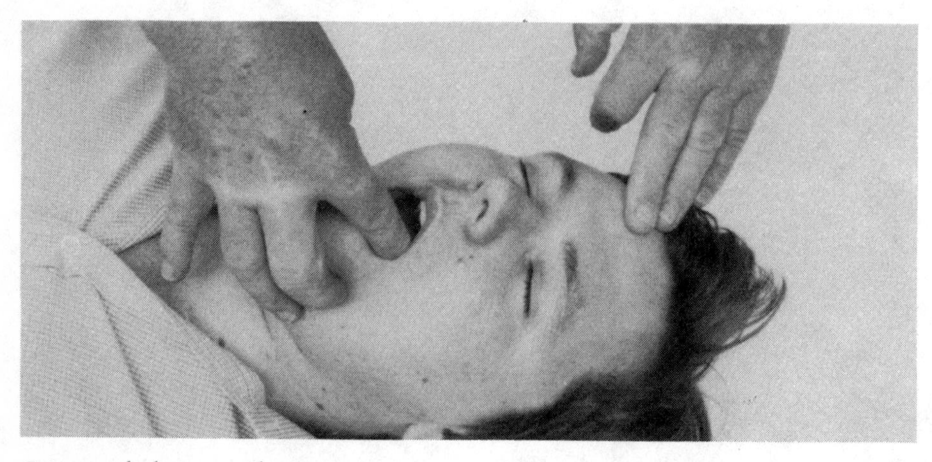

Open and clear mouth.

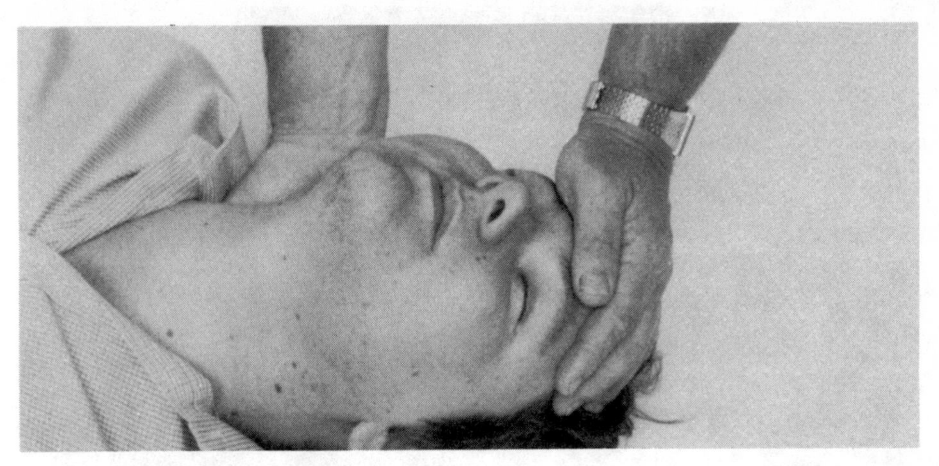

Tilt head backwards to open airway. Maintain this position throughout.

Support jaw as shown.

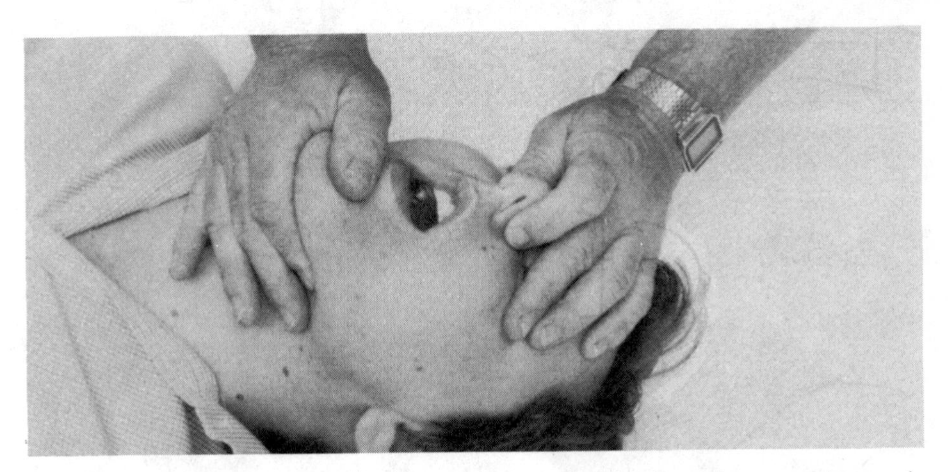

Kneel beside casualty. While keeping his head tilted backwards, open his mouth and pinch his nose.

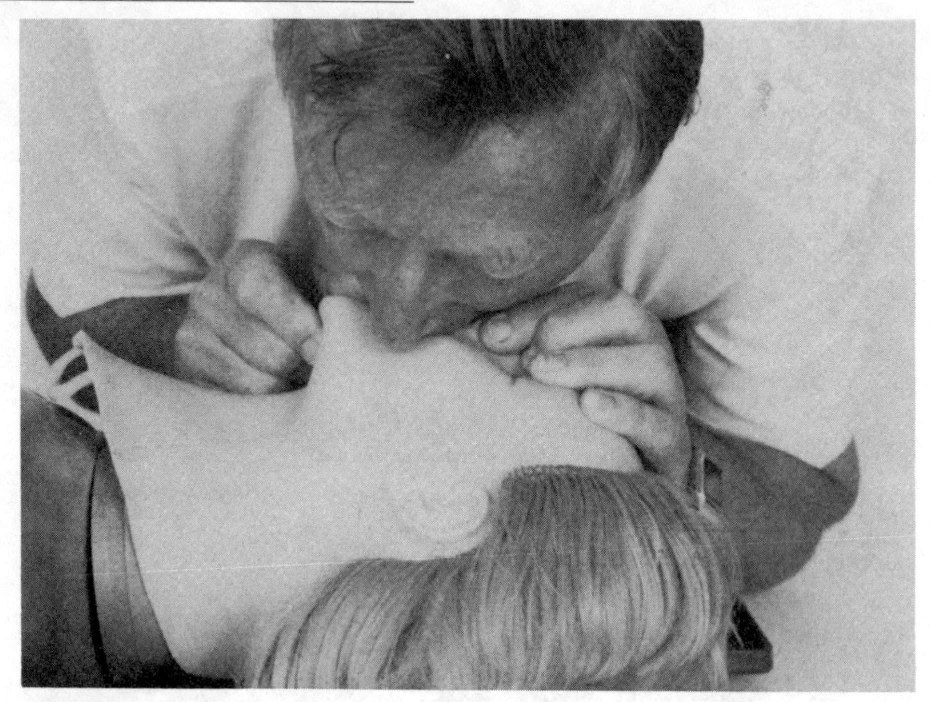

Open your mouth, take a deep breath. Seal his mouth with yours and breathe out firmly into it.

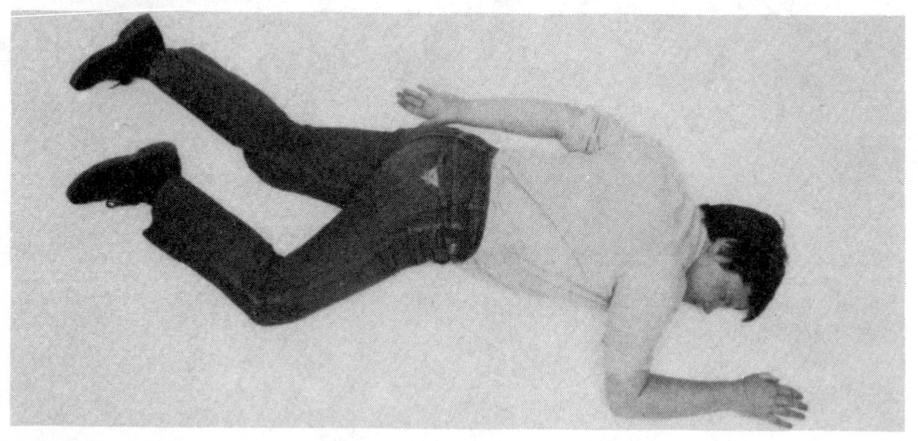

Casualty's chest should rise. Then remove your mouth and let his chest fall. If chest does not rise, check head is tilted backwards sufficiently. Continue at a rate of 12 times a minute until the casualty is breathing by himself. Place casualty in the recovery position as shown.

(2) **UNCONSCIOUSNESS**

Place casualty in the recovery position as shown.

(3) SEVERE BLEEDING

Control by direct pressure (using fingers and thumb) on the bleeding point. Apply a dressing. Raising the bleeding limb (unless it is broken) will help reduce the flow of blood.

OTHER CONDITIONS

(4) SUSPECTED BROKEN BONES

Do not move the casualty unless he is in a position which exposes him to immediate danger.

(5) BURNS

BURNS AND SCALDS

Do not remove clothing sticking to the burns or scalds or burst blisters. If burns and scalds are small, flush with plenty of clean, cool water before applying a sterilised dressing. If burns are large or deep, wash your hands, apply a dry sterile dressing and send to hospital.

CHEMICAL BURNS

Avoid contaminating yourself with the chemical.

Remove any contaminated clothing which is not stuck to the skin. Flush with plenty of clean, cool water for 10–15 minutes. Apply a sterilised dressing to exposed, damaged skin and send to hospital.

(6) EYES

Loose foreign bodies in the eye: Wash out eye with clean, cool water.

Chemical in the eye: Wash out the open eye continuously with clean, cool water for 10–15 minutes.

People with eye injuries should be sent to hospital with the eye covered with an eye pad.

(7) ELECTRIC SHOCK

Do not touch the casualty until the current is switched off. If the current cannot be switched off, stand on some dry insulating material and use a wooden or plastic implement to free the casualty from the electrical source. If breathing has stopped, start mouth to mouth breathing and continue until casualty starts to breathe by himself or until professional help arrives.

(8) **GASSING**

Use suitable protective equipment.

Move casualty to fresh air.

If breathing has stopped, start mouth to mouth breathing and continue until casualty starts to breathe by himself or until professional help arrives. Send to hospital with a note of the gas involved.

(9) **MINOR INJURIES**

Casualties with minor injuries of a sort they would attend to themselves if at home may wash their hands and apply a small sterilised dressing from the first-aid box.

(10) **RECORD KEEPING**

An entry of each case dealt with must be made in the accident book.

(11) **FIRST-AID MATERIALS**

Articles used from the first-aid box should be replaced as soon as possible.

MANAGEMENT ACTION CHECKLIST 12

The Health and Safety (First-Aid) Regulations 1981

Checkpoints	Action required Yes	No	Action by
Are you satisfied that the company first-aid arrangements are adequate, taking account of all the relevant factors?			
Do you have the first-aiders and equipment in place to reflect your declared provision?			
Have you communicated your first-aid arrangements to employees using the following media: (a) notice-boards; (b) safety policy; (c) house telephone directory; and (d) induction (for new starters)? You should be using at least (a), (b) and (d).			
Is there an assigned employee responsible for all administrative matters relating to first-aiders, including but not limited to: recruitment, training, records, ensuring complement is maintained, waiting list, certificates framed, regular meetings?			
Do all your staff know who the 'appointed person' is?			
Do your first-aid boxes contain only the prescribed contents, and are they regularly checked and replenished?			
Do you have a first-aid room? if yes, is it fitted out and equipped properly?			
Is your first-aid or rest room cleaned each day within the cleaning specification?			
Who checks the treatment book(s) to ensure that they are being maintained properly? It is a statutory duty.			

13

MISCELLANEOUS REGULATIONS AND REQUIREMENTS

This miscellaneous chapter is not and cannot be an exhaustive guide to every health and safety regulation. Instead it concentrates on the key regulations: that is, those that affect most workplaces. It includes some minor regulations of general application, for example the Health and Safety Information for Employees Regulations, and finally some others which apply only to certain premises, for example those with cooling towers or evaporative condensers.

The information provided for each regulation included is not exhaustive; rather it is a guide to the essential requirements. The following subjects and relevant regulations are covered:

- Electricity
- Employers' liability insurance
- Food hygiene
- Health and safety information for employees
- Lifts
- Legionnaires' disease/cooling towers
- Noise
- Pressure systems
- Safety signs
- Thermometers

ELECTRICITY – THE ELECTRICITY AT WORK REGULATIONS 1989

These regulations bring together, for the first time, *every* aspect of the electrical system in a premises; they place duties not only upon employers, but upon employees and the self-employed.

The difficulty is that electricity and its management are matters which many managers – and employees – feel inadequately and insufficiently trained to deal with. While this does create problems, an understanding of the checks and balances which can be carried out to demonstrate a responsible and caring approach is important.

CHECKLISTS

There are three checklists which will serve as management prompts for three different aspects of electrical work.

1. Fixed electrical systems

The important message in the checklist given in Table 13.1 is that although the work itself might be undertaken and supervised by others, the local manager *must* be thoroughly briefed. He or she must also have a basic knowledge of the reason(s) for such work and of what dangers, if any, may be created by it.

Table 13.1 Checklist for fixed electrical systems

1. Is the design of new, or extensions to existing, electrical systems carried out by staff with the appropriate technical knowledge and experience, and an understanding of the 16th edition of the IEE Requirements for Electrical Installation (BS 7671) of the Institution of Electrical Engineers (IEE), British Standards and HSE guidance?
2. Are all items of electrical equipment forming part of the system and other system components, specified to the appropriate British or other recognized standards?
3. Have all the items of electrical equipment and components been selected to take account of the environment in which the electrical system is to be installed?
4. Is all new electrical installation work inspected and tested prior to handover or putting into service?
5. Is all equipment clearly labelled, particularly switchgear and fuseboards, for circuits identification purposes?
6. Are circuit diagrams and/or plans kept to provide a comprehensive record of all electrical systems, and do arrangements exist for updating these following system modifications?
7. Are all electrical systems periodically inspected and tested and appropriate records kept?

Source: P. C. Buck and E. Hooper (1992), *Electrical Safety at Work – A Guide to Regulations and Safe Practice*, Boreham Wood: Paramount Publishing.

2. Portable electrical tools and appliances

The checklist given in Table 13.2 covers the requirements under this section. Only parts of this list will apply to most managers.

Table 13.2 Checklist for portable electrical tools and appliances

1. Are all portable electric tools and electrical appliances periodically inspected and/or tested to ensure that they are still fully serviceable, and appropriate records kept?
2. Are these inspections and tests carried out by competent personnel, working in accordance with specified procedures?
3. Is appropriate equipment available, and used, to carry out the necessary testing?

Source: P. C. Buck and E. Hooper (1992), *Electrical Safety at Work – A Guide to Regulations and Safe Practice*, Boreham Wood: Paramount Publishing.

3. Systems of work involving electricity

Again, not all items in the checklist given in Table 13.3 may be applicable to all managers. Nevertheless it should be studied with care and applied as appropriate.

The following guidelines are intended to amplify some of the items appearing in the three checklists.

BASIC ELECTRICAL WORK

Many of the things we do at work almost routinely, and without great consideration of danger, now fall within the scope of the 1989 Regulations. For example:

- Changing light bulbs, fluorescent tubes and their starter units.
- Changing fuses (cartridges and fuse wire).
- Plugging in and out of sockets.
- Creating new circuitry by using extension leads.
- Switching on and off at switches and at main isolators.
- Withdrawing fuses when minor electrical work is carried out.
- Moving the position of portable electric equipment.
- Fitting 13 amp plugs – to do this properly requires skill and understanding.
- Purchase and use of portable electric devices.

While the changing of light bulbs is a basic task, the fact remains that management have to satisfy themselves that the person(s) carrying out such tasks is 'competent'. For example, do they understand that the local switch must be off even for the simple 60 watt bulb change?

COMPETENCE

The difficulty in ensuring the competence of electrical contractors, or compliance with standards in respect to electrical equipment, is understood. The

Table 13.3 Checklist for systems of work involving electricity

1. Are there written and understood procedures for undertaking work on electrical systems and equipment, on the basis that danger is removed by disconnection of the electricity supply?
2. Do these procedures specify the requirements for the necessary safe systems of work, including the arrangements to ensure that the isolation is secure, and for proving dead at the point of work?
3. Is there a clear management policy setting down the justified circumstances when live working is permitted?
4. When live working is justified and is carried out, are there properly documented work procedures to prevent injury?
5. Do the procedures for live work specify the requirements for the necessary safe systems of work, including the provision and use of specialist equipment such as insulated tools, rubber gloves, rubber mats and screening material?
6. Do those staff required to undertake electrical work first receive appropriate training, whether this be for work on de-energized or live systems?
7. Do staff receive refresher or additional training when required?
8. Are checks undertaken, on a regular basis, to confirm the competency (technical knowledge and experience) of staff engaged on electrical work, and that this is appropriate to the type of electrical system and work to be undertaken?
9. Are records kept of all training and assessments?
10. Are staff given adequate instructions for all work, including information on relevant hazards and precautions to be taken?
11. During the course of the work, do staff receive adequate supervision?
12. Do you, as a manager in control, understand your duties under the regulations?

Source: P. C. Buck and E. Hooper (1992), *Electrical Safety at Work – A Guide to Regulations and Safe Practice*, Boreham Wood: Paramount Publishing.

following guidance, if followed, will demonstrate management concern to ensure that safety in electricity is maintained.

Equipment

Purchase goods of good quality and design. Indicators are goods which have been approved by the British Electrotechnical Approvals Board (BEAB), British Standards Institution (BSI Kite mark) or other reputable bodies. Inexpensive 'offers' may *not* meet these criteria.

Contractors

Employ only qualified electricians to install wiring and equipment. Choose a contractor who is a registered member of the National Inspection Council for Electrical Installation Contracting (NICEIC). To obtain further details of this inspecting body, telephone 071 582 7746.

INSPECTIONS AND RECORD KEEPING

All portable appliances and 'pluggable' equipment should be inspected by a competent electrician at an agreed frequency, and records of such inspections kept in an 'electricity log'. This log can be an exercise book laid out to record such inspections, and should contain details of *all* electrical work carried out in the location.

It was originally considered necessary to inspect portable/pluggable equipment at least once every year. This gave rise to widespread criticism that the requirement was too onerous and prohibitively expensive. A more pragmatic approach is clearly necessary.

Initially an annual check – perhaps more often where wear and tear is high – should be carried out in order to establish some base data. The result of the first two or three tests will indicate what the ongoing frequencies should be. The intervals between some checks might have to remain short, with longer periods for equipment that has a low incidence, or possibly no incidence, of failure. There will be differences, for example, between portable equipment that is moved about frequently, such as overhead and video projectors. Sometimes this type of equipment is transported in cars and is therefore subject to considerable wear. Conversely equipment which remains in one office for the whole of its life is less likely to cause problems.

The employee or contractor carrying out the regular inspections of portable equipment should develop a unique code of endorsement of the items inspected that can be understood by all staff, who will then be responsible for ensuring that every item of electrical equipment which they use bears an 'in date' certification. It will be part of the location procedures (safe system of work) that no employee may use an item of electrical equipment which is not 'in date', and must immediately bring that item to the attention of their manager.

Each location must maintain – and keep up to date – a schedule of *all* the electrical equipment in the location, to ensure that it is reviewed by the electrical contractor. Generally speaking, provided that new acquisitions comply with the above section on 'Equipment', these items can await further inspection until the next annual review. However, if managers are in any doubt, they should arrange a one-off inspection and certification.

Managers cannot allow *any* private or personal item of electrical equipment to be used in their locations. Such equipment is often discarded at home, and brought to work to be used for a little longer. These items are usually old when brought onto the premises, get the worst kind of treatment, and are most prone to causing accidents.

Allowing the use of such equipment exposes individual managers and the company to considerable risk and the possibility of legal penalties.

SWITCHGEAR

Electrical cupboards are *not* store cupboards. They should be kept locked shut, scrupulously clean, and have the electricity 'Danger' symbol on the outer door, together with a standard safety sign 'To be kept locked shut'. Inside such rooms, in cases where live working occurs, there should be a standard electric shock treatment poster.

THE FOOD SAFETY ACT 1990 AND THE FOOD HYGIENE (GENERAL) REGULATIONS 1970

This legislation is intended to gather all food-related activities within the legislative fold, including those concerned with our food growing and production.

In so far as businesses are concerned, there can be complications in determining what falls within the legislation, and what does not. Generally speaking there is no intention to control food-related matters where there is no gain or profit. Therefore, where staff contribute to or provide the ingredients for coffee/tea making and so on, this does not fall within the regulations. Staff canteens or restaurants do.

There is a requirement formally to notify the existence of registerable premises. The onus is upon the person carrying on the food business to make the registration, and no application for registration can be refused. The object here is to ensure that the appropriate authorities are aware of every food premises, and from that information they can then direct their attention on premises (including vehicles) likely to be operating in an unsafe/unhygienic manner.

The list that follows summarizes the key requirements or highlights matters which inspectors have commented on or required during visits to catering premises or departments around the country:

- Catering staff must receive training in catering hygiene. Precise standards for this training have not yet been issued but it is considered good practice to provide it and most responsible catering contractors do so. In some areas, local environmental health departments are themselves providing training sessions. The advantages in catering staff attending such courses will be obvious.
- Food handlers must not handle money, especially coin.
- Handwashing facilities must be separate and distinct from sinks for food preparation. These facilities are often not used at all or are commandeered for other purposes. The separate identity of the hand-washing facilities should be obvious, with clear notices and containing only the usual requisites (e.g. liquid soap, nailbrush, drying facilities).
- It is not sufficient simply to say that fridge/freezer temperatures are noted. There should be documentary evidence, with clear indication of

the temperature parameters or optimum temperature on the equipment and in the check log.

- There must be stringent hygiene standards for food handlers (e.g. obviously no smoking in or about food areas; all wounds/cuts to be kept covered with blue, easily identified, waterproof plasters; handlers to report illness/disease immediately).
- Accident books to be kept in good condition, and complete and up to date.
- Separate sanitary conveniences to be sited away from kitchens.
- Facilities must be provided for changing into/out of kitchen uniform. Note that these are not to be just the backs of chairs in the manager's office. Separate changing facilities for both sexes will be required in 1996 in accordance with the Workplace (Health, Safety and Welfare) Regulations 1992 (HSW) (see Chapter 17).
- Lighting and ventilation must be adequate.
- No food to be stored or placed on floors.
- Separation of cooked from uncooked food must be maintained.
- Impervious surfaces must be used for the storage and preparation of food.
- Kitchens must be kept scrupulously clean and regularly cleaned, with no accumulation of spilt material, and any spillages to be cleaned up immediately.
- Any structural abrasions, imperfections, breaches, or breaks in walls, floors or ceilings, to receive immediate attention.

THE LIFTING PLANT AND EQUIPMENT REGULATIONS 1992

The requirement for a regular (six-monthly) examination of lifts, including passenger lifts, remains a statutory duty in accordance with the above regulations, which amended earlier complex regulations on this and related subjects.

The emphasis is likely to be on what local officials require as opposed to a blanket rule across the country. For the time being it is recommended that the old approach is followed, namely, that each lift is inspected every six months and a report issued. The inspection will be by a qualified engineer nominated by the company's insurance carrier.

A report will be issued following inspections. This will classify the lift as follows:

1. With no observations or requirements – a 'clean sheet'.
2. Report with some observations.
3. Report with observations and/or positive requirement.

If in the engineer's opinion, the lift as seen has implicit dangers in use, he may either ask for it to be taken out of service, or, if kept in service, it must have some immediate improvements carried out. In such cases, the local statutory authority will be informed by the inspecting engineer.

THE NOTIFICATION OF COOLING TOWERS AND EVAPORATIVE CONDENSERS REGULATIONS 1992

These regulations were made so that local authorities could obtain some information on the extent of the risk of legionnaires' disease outbreaks in their areas. The requirement is to register with the local authority any 'notifiable devices' which are defined as 'all cooling towers and/or evaporative condensers – except where the water is not exposed to air and the water and electrical supplies are not connected'.

The following notes on legionnaires' disease do not form part of the above regulations.

Risks

Legionellae exist everywhere in very small quantities, and represent no threat unless/until prevailing conditions favour proliferation.

Optimum conditions for proliferation

- Presence of nutrients – sludge, scale, rust, algae.
- Water temperatures in range 20–45°C. Body temperature (37°C) is optimal.
- Does not multiply below 20°C.
- Cannot survive a continuous temperature of 60°C.
- Are killed instantly at 70°C.

Note: Danger exists when droplets are dispersed, as, for example, in aerosol-like action.

Areas of danger

- Air conditioning system cooling towers.
- Industrial cooling systems.
- Whirlpools, spas, showers, fire sprinkler systems.

How to eliminate

Better design, construction, operation and maintenance of water installations.

NOISE

HEARING LOSS FROM WORK ACTIVITY – THE LONG AWAKENING

It has taken a long time for the deleterious effects of noise at work to be fully recognized, and even longer for legislation to be enacted to protect workers from excessive workplace noise, despite the fact that the dangers were first perceived almost 200 years ago! Deafness arising from occupational noise was diagnosed among smiths, boilermakers and braziers at the beginning of the last century, well over 150 years before positive steps were taken to deal with occupational deafness.

It was not until 1963 that the then Inspectorate of Factories published an advisory booklet, *Noise and the Worker.* This booklet was the first official statement on the danger of noise to the hearing of workers, and effectively gave notice to employers that they should have regard to the dangers which their noisy operations might create. This was followed nine years later (1972) by the publication of a code of practice, *Code of practice for reducing the exposure of employed persons to noise.* The Health and Safety at Work Act 1974, although making no specific reference to noise, none the less strengthened the 1972 code of practice. Section 2 of the Act, with its requirement for all employers to 'provide and maintain a working environment for his employees that is, so far as is reasonably practicable, safe, without risks to health', and so on, served as a means to enforce the code more effectively.

Even so, the extent to which statutory authorities used the legislation to bring about improvements may be gauged by the fact that there were only 20 improvement notices relating to noise issued in 1975, and this had only risen to 135 in 1984. The 1984 figure represented only 2 per cent of all improvement orders issued, yet it was recognized that the incidence of exposure to noise, particularly in manufacturing industry, was high. At this time it was estimated that approximately half of all workers were exposed to 80dB(A), 10 per cent to 90 dB(A) and about 2 per cent to 100 dB(A). In addition to manufacturing industry, high levels of noise exposure were common to construction, transport, shipping, agriculture, quarrying, forestry and entertainment.

Where workers have brought actions for damages in respect to hearing impairment or loss, the damages awarded indicate that courts have not fully appreciated the seriousness of hearing disability: *Berry* v. *Stone Manganese Marine Ltd* [1972] – £2,500 halved due to time limitation; *Heslop* v. *Metalock (Great Britain) Ltd* [1981] – £7,750; *O'Shea* v. *Kimberley-Clark Ltd* [1982] – £7,490 (tinnitus); *Tripp* v. *MOD* [1982] – £7,500. In a more recent case account was taken, among other factors, of the social handicap suffered: *Bixby, Case, Fry and Elliott* v. *Ford Motor Group* (1990), unreported.

During the past 40 years, there have been changes in the pattern of noise creation. With the decline of heavy industries, and as a result of technological innovation, many traditional sources of industrial noise have declined or disappeared altogether. Similarly, soundproofing improvements to modern office machinery have had an effect. The age of the typing pool has passed into history – or almost – and the electronic equipment which has replaced the old manual typewriter is certainly less noisy.

Unfortunately these changes in workplace noise have not been matched with similar noise reduction elsewhere, and there is little doubt that we are all exposed to noise levels outside work that are greater than those encountered by earlier generations. The factory hooter and siren may have been silenced, but the cacophony of noise created by motor cycles, the volume of traffic generally, and, not least, by modern forms of music, have created ambient noise levels that far outweigh those of earlier times. Moreover, the spread of this noise is universal, and there is often no escape from it. The efforts of the factory owner to mitigate noise generation will be of little real value to the employee who rides a noisy motor cycle, wears a Walkman speaker during most of his waking life, and only takes it off to spend four or five hours at a disco!

The UK's legislative approach to the noise problem has been tardy to say the least, and the legislation has arrived at a point in time when much hearing damage has already been done, and where the lifestyle of many – particularly the young – is causing hearing damage far more insidious than the noise to which they are exposed at the workplace. This gives rise to the possibility that employees suffering from tinnitus (ringing in the ears), or deafness, may have brought this misfortune on themselves as a consequence of what they do in their leisure time, and not as a result of work-related noise.

Employers, particularly those who are under a statutory obligation to take measures to overcome noise at the 'action levels' required by the Noise at Work Regulations 1989, should be alert to the possibility of compensation claims from employees whose loss of hearing is more likely to be self-inflicted than caused by the noise to which they are exposed while at work. This is not to say that noisy workplaces do not constitute a risk to hearing, simply that such noise may only be one of the contributory factors. If an employer cannot dictate the personal lifestyles of his workers, he can at least protect himself from unjustified claims for compensation for hearing loss by initiating a 'hearing conservation programme'. This will include carrying out hearing tests for staff when they commence work – or *before they commence work in noisy operations* – and at regular intervals thereafter. This will not provide an employer with a complete defence against unjustified claims – for example if an employee engages in a noisy personal lifestyle *after* commencing noisy work – but it will prevent many abuses.

All businesses, including those which are not inherently noisy, are affected by the Noise at Work Regulations 1989. These regulations require *every* employer to reduce the risk of hearing damage to the lowest level reasonably practicable. This means that even the traditionally quieter workplaces such as offices have to consider how to reduce ambient noise levels – even when these are far below the stipulated 'action level' thresholds described later in this chapter.

The importance of the need to consider noise in the office has been heightened by the new display screen equipment regulations 1993 (see Chapter 16). Display screen equipment (VDU) 'users', when questioned as part of the 'assessment' exercise demanded by these regulations, frequently complain that they experience 'nuisance' noise, sometimes from air conditioning systems, but more often from adjacent office machinery such as printers and copiers. Apart from the debilitating effect of this noise, business operations are actually interrupted because the noise prevents staff from using the telephone, and sometimes from even conversing properly.

Experts recognize that noise is a significant contributor to stress and other illnesses, and should be reduced. Unfortunately non-workplace noise, although covered by the Control of Pollution Act, is unlikely to be seriously addressed while there are few officials to monitor and police noise, and where litigation is often left to the person or persons who are the victims of unwanted noise.

It will be seen from the legislation paragraphs of this chapter, that there is a requirement to address peak noise – noise of perhaps very short duration, which none the less poses a threat to hearing where this occurs in the workplace. Outside the workplace, it is frequently the peak noise which creates the greatest problem – a motor cycle roaring through the high street causing all conversation to stop, frightening children and the elderly, and bringing normal life to a standstill. Unless the environmental health officer happens to be present at the time, there is little chance of effective action being taken.

There are two distinct legislative instruments relating to workplace-generated noise:

1. The Environmental Protection Act 1990 This Act deals, *inter alia*, with noise emitted by a business which is either a nuisance or poses a health risk to persons not employed by the business concerned. If such noise is not abated, fines on conviction can be up to £20,000, plus £2,000 for every day that the nuisance or risk continues. The same legislation covers 'domestic' noise nuisance (e.g. music amplification); in such cases the maximum fines in magistrates' courts are £2,000 plus £200 for every day that the nuisance continues.

2. The Noise at Work Regulations 1989 These regulations relate to noise affecting persons *inside* the employer's premises.

Although there is a general duty for all employers to reduce the risk of hearing damage to the 'lowest level reasonably practicable', there are definitive requirements where workplace noise reaches certain 'action levels'.

ACTION LEVELS

1st action level

Where the noise levels are between 85 and 90 dB(A).

2nd action level

When noise levels exceed 90 dB(A).

Peak action level (sometimes referred to as the 3rd action level)

Where the peak sound pressure is at or above 200 pascals.

EMPLOYER'S DUTIES

- To reduce the risk of hearing damage to the lowest level reasonably practicable.
- To initiate a noise assessment when there is a likelihood that noise is approaching or has reached the 1st action level. If colleagues cannot converse with each other when two metres apart without the need to shout, this warns of a need to carry out or commission a noise survey.
- To ensure assessment is carried out by a competent person.
- Record of assessment to be kept until a new one is made.

- Noise reduction must be carried out by means other than hearing protectors if noise reaches 2nd or 3rd action levels.
- To provide information, instruction and training about risks, how to minimize them, how to obtain hearing protectors, and about the employer's obligations to his employees.
- To mark 'ear protection zones' with notices and so forth if the noise levels reach the 2nd or 'peak' action levels (see concluding paragraph).
- To provide ear protectors: (a) to employees that ask for them when noise reaches the 1st action level; and (b) automatically if noise reaches 2nd or 'peak' action levels.
- To maintain and repair ear protectors.
- To ensure that ear protectors are worn by all employees working in areas where 2nd or 'peak' action levels exist.
- To ensure that ear protectors selected meet the requirements of the task, offer appropriate protection, *and* are suited to the wearer.

EMPLOYEE'S DUTIES

So far as is practicable:

- To use ear protectors for 2nd and 'peak' action levels.
- To use any other equipment provided by the employer, and report any defect to his or her employer at *all* action levels.

MACHINE MAKERS' AND SUPPLIERS' DUTIES

To provide information on noise likely to be generated.

Table 13.4 summarizes all the above duties.

CONCLUSION

For cultural and historic reasons many workers are uncomfortable with what they perceive to be displays of overt concern for their health and safety, and failure to wear ear protectors is often a manifestation of this attitude.

It is therefore imperative that management set an example in this respect, even when they are simply passing through an ear protection zone. Nothing does more to disabuse workers of the importance of ear protection than to see management ignoring the rules, or worse, for management to be seen accompanying visitors through the area when none of the party are complying.

When this practice occurs, the reason is more likely to be the concern not to delay or inconvenience visitors – who may be customers or prospects – than a disregard for safety. In fact, this concern might have the opposite effect to that intended. Customers and prospects are becoming increasingly concerned that they deal with safety-conscious companies, and if they see that potential suppliers are paying little attention to ear protection – and therefore, by implication, to safety as a whole – this could result in loss of business.

Table 13.4 Summary of the duties of all parties in relation to noise in the workplace

Action required where $L_{EP.d}$ is likely to be:[1]	Below 85dB(A)	85dB(A) First AL	90dB(A) Second AL[2]
Employer's duties			
General duty to reduce risk			
Risk of hearing damage to be reduced to the lowest level reasonably practicable (Regulation 6)	■	■	■
Assessment of noise exposure			
Noise assessments to be made by a competent person (Regulation 4)		■	■
Record of assessments to be kept until a new one is made (Regulation 5)		■	■
Noise reduction			
Reduce exposure to noise as far as is reasonably practicable by means other than ear protectors (Regulation 7)			■
Provision of information to workers			
Provide adequate information, instruction and training about risks to hearing, what employees should do to minimize risk, how they can obtain ear protectors if they are exposed to between 85 and 90 dB(A), and their obligations under the regulations (Regulation 11)		■	■
Mark ear protection zones with notices, so far as reasonably practicable (Regulation 9)			■
Ear protectors			
Ensure so far as is practicable that protectors are:			
• provided to employees who ask for them (Reg 8(1))		■	
• provided to all exposed (Reg 8(2))			■
• maintained and repaired (Reg 10(1)(b))		■	■
• used by all exposed (Reg 10(1)(a))			■
Ensure so far as reasonably practicable that all who go into a marked ear protection zone use ear protectors (Regulation 9(1)(b))			■
Maintenance and use of equipment			
Ensure so far as is practicable that:			
• all equipment provided under the regulations is used, except for the ear protectors provided between 85 and 90 dB(A) (Regulation 10(1)(a))		■	■
• ensure all equipment is maintained (Regulation 10(1)(b))		■	■

Cont'd

Table 13.4 (continued)

Action required where $L_{EP,d}$ is likely to be:[1]	Below 85dB(A)	85dB(A) First AL	90dB(A) Second AL[2]
Employee's Duties			
Use of equipment			
So far as practicable:			
• use ear protectors (Regulation 10(2))			■
• use any other protective equipment (Regulation 10(2))		■	■
• report any defects discovered to employer (Regulation 10(2))		■	■
Machine Makers' and Suppliers' Duties			
Provision of information			
Provide information on the noise likely to be generated (Regulation 12)		■	■

Notes: [1]The dB(A) action levels are values of daily personnel exposure ($L_{EP,d}$)

[2]All the actions indicated at 90 dB(A) are also required where the peak sound pressure is at or above 200 Pa (140 dB re 20 μPa).

[3]This requirement applies to all who enter the zones, even if they do not stay long enough to receive an exposure of 90 dB(A) ($L_{EP,d}$).

PRESSURE SYSTEMS AND TRANSPORTABLE GAS CONTAINERS REGULATIONS 1989

These regulations will only affect businesses who have pressure systems where the contents are 0.5 bar or more above atmospheric pressure, or where there is any steam present.

The principal features of these regulations are that they gather up and improve a number of separate and outdated regulations, and they apply to pressure systems in their entirety: that is, they include all the associated pipework, not just the vessels themselves as was formerly the case.

Applicable systems need to be reviewed by a competent person, and must be the subject of a 'written scheme for their subsequent maintenance and testing'. Previously testing had to be carried out at prescribed intervals, and within an established procedure.

For advice about any systems which might be within the ambit of these regulations, the current insurance examiner of the system should be consulted.

WORKPLACE (HEALTH, SAFETY AND WELFARE) REGULATIONS (HSW) 1992

THERMOMETERS

Regulation 7 of HSW deals with temperatures, and, under paragraph 57, requires thermometers to be available at a convenient distance from every part of the workplace for persons to verify temperatures, although they need not be provided in each workroom.

Notwithstanding that HSW is one of the six EU-driven sets of regulations allowing a transition period for compliance, this will not be allowed in respect to thermometers, a requirement for which has always existed for offices in accordance with the Offices, Shops and Railways Premises Act.

Although the practice of displaying thermometers has declined in recent years, inspectors were beginning to reinforce the need for their provision last year – well before this requirement was in turn reinforced by HSW.

The HSW requirements are covered in full in Chapter 17.

HEALTH AND SAFETY INFORMATION FOR EMPLOYEES REGULATIONS 1989

Under these regulations a statutory poster must be displayed in all workplaces. Before being put up, the poster must be completed to show the address of the local enforcing authority and office of EMAS (Employment Medical Advisory Service). It is an offence:

1. Not to display the notice (unless every employee is issued with a pamphlet version of it. This would be expensive and unreliable re leavers and new starters.)
2. To display the poster without completing the two address boxes referred to above.

EMPLOYERS' LIABILITY (COMPULSORY INSURANCE) ACT 1969

This statute requires the display of a copy of the current certificate of this insurance in such a way as to enable all employees to verify that such a policy exists and is in date.

THE SAFETY SIGNS REGULATIONS 1980

These regulations were enacted to ensure harmonization of safety-type signs erected within and around business premises across the EU. Road signs had previously been Europeanized here.

The rationale for these signs is that they conform to a different colour and background according to their purpose. The main (statutory) sign groupings are as follows:

1. *Warning* Yellow background triangle with black letters and/or symbols.
2. *Mandatory – must do* White letters/symbols on a blue background.
3. *Prohibition* White lettering/symbols on a red background.
4. *Safe condition – the safe way* White lettering/symbols on a green background.
5. *Fire fighting* White letters/symbols on a red background.

A sign which falls within any of these five groupings must conform to the above standards. It is an offence not to comply. The problem with signs that do not conform is that it will take longer for employees to become accustomed to the meaning and relevance of these sign standards. This problem will be exacerbated as more signs are brought within the scope of these regulations.

In the case of a message which needs to combine more than one of the five groupings, it is permissible to use a multipurpose sign providing that the colour combinations are complied with.

SOME PROBLEMS WITH SAFETY SIGNS

- Old signs not being changed to conform with the above. The deadline for old signs to be changed was 1986. A common breach is where fire emergency exit signs are still coloured red/white.
- *Ad hoc* safety signs which are not converted to properly made signs conforming with the above standards.
- Using local wording for signs for which an official version exists.
- Not using the correct sign for the desired purpose. For example, using a 'No pedestrian access' message on a yellow 'warning' sign instead of a red 'prohibition' sign.
- Confusing signs which relate to exit and escape with 'fire fighting' signs.
- Manufactured signs – in particular 'Exit' and 'Fire exit' – where the manufacturer has used the correct colour combination, but transposed them. For example, 'Fire exit' with the correct green/white colours, but transposed so that the lettering is green and the background white. Not only wrong, but reduces visibility of the sign.

FURTHER READING

Electricity

Buck, P. C. and Hooper, E. (1992), *Electrical Safety at Work: A Guide to Regulations and Safe Practice*, Boreham Wood: Paramount Publishing.

Health and Safety Executive, *The Electricity at Work Regulations 1989: An open learning course*, London: HSE Books. Priced at £14.00.

Health and Safety Executive, *Memorandum of guidance on the Electricity at Work Regulations 1989*, HSE safety series booklet no. HS(R)25, London: HSE Books. Priced at £4.00.
HSE Books also publish a large series of GN publications on electricity.

Food hygiene

MAFF (1992), *The Food Safety Act 1990 and you – a guide for the food industry*, no. PB0351, (revised 1992), London: Ministry of Agriculture, Fisheries and Food (MAFF).
MAFF (1990), *The Food Safety Act 1990 and you – a guide for caterers and their employees*, no. PB0370, (revised July 1992), London: Ministry of Agriculture, Fisheries and Food (MAFF).
Both publications are available free from Food Sense, London, SE99 7TT.

Lifting plant and equipment

Health and Safety Executive, *A guide to the Lifting Plant and Equipment (Records of Test and Examination, etc.) Regulations 1992*, legislation (L) series, no. L20, London: HSE Books. Priced at £2.50.

Legionnaires' disease

Health and Safety Executive, *Approved Code of Practice (ACOP) for the prevention or control of legionellosis (including legionnaires' disease)*, HSE booklet no. L8, London: HSE Books. Priced at £3.00.
Health and Safety Executive, *The control of legionellosis including legionnaires' disease*, HSE booklet no. HS(G)70, London: HSE Books.
Health and Safety Executive, *Legionnaires' disease*, rev. edn, guidance leaflet no. IAC(L)27, London: HSE Books. Available free of charge from HSE Information Centre, Broad Lane, Sheffield S3 7HQ (telephone: 0742 892345).

Noise

Health and Safety Executive (1989), *Noise at Work: Noise guides*, No. 1, *Legal duties of employers to prevent damage to hearing*; No. 2, *Legal duties of designers, manufacturers, importers and suppliers to prevent damage to hearing*, London: HSE Books. Priced at £3.00 each.
Health and Safety Executive, *The Noise at Work Regulations: A brief guide to the requirements for controlling noise at work*, no. HSE IND(G)75(L), rev. edn, London: HSE Books. Available free of charge.
Health and Safety Executive, *Noise at Work: Advice for employees*, no. HSE IND(G)99(L), rev. edn, London: HSE Books. Available free of charge.

Pressure systems

Health and Safety Executive, *ACOP: Safety of pressure systems*, London: HSE Books. Priced at £4.50.
Health and Safety Executive, *ACOP: Safety of transportable gas containers*, London: HSE Books. Priced at £4.00.
Health and Safety Executive, *A guide to the Pressure Systems and Transportable*

Gas Containers Regulations 1989, London: HSE Books. Priced at £4.50.

Health and Safety Executive, *Introducing Competent Persons: Pressure Systems and Transportable Gas Containers Regulations 1989*, no. IND(S)29(L), London: HSE Books. Available free of charge from all HSE public enquiry points.

MANAGEMENT ACTION CHECKLIST 13
Miscellaneous Regulations and Requirements

Checkpoints	Action required Yes	No	Action by
Electricity			
What is the portable electrical equipment inspection/testing programme in the company?			
Are details published, and do all staff understand that they may not use – and must report – any item not 'in date'?			
What arrangements exist to ensure new equipment is added to the register?			
Is there any personal portable/pluggable electrical equipment still on company premises? If yes, how will the employer's statutory duty re electricity be discharged?			
How do you/your company rate against checklists 1, 2 and 3 in the electrical section of this chapter?			
Food safety/hygiene			
Is there a registerable catering facility in the company? If yes, has it been registered?			
If catering is staffed by company employees, they must receive appropriate training. Have they received such training within the past 18 months?			
If catering is contracted, is there close liaison with the caterer's management?			
As occupiers you have the right to inspect the catering facility. Is this regularly done?			
Lifting plant			
Is there lifting plant, including passenger lifts on company premises? If so, is there a regular engineering inspection carried out?			
Who manages the inspection?			
Who keeps the documentation?			
What is the procedure for implementing the recommendations/requirements of the inspecting engineer?			

Reproduced from *What Every Manager Needs to Know About Health and Safety* by Ron Akass, Gower, Aldershot, 1995.

Checkpoints	Action required Yes	No	Action by
What checks and balances exist to ensure that all the above is being done properly?			
Cooling towers/legionnaires' disease			
Are there any 'notifiable' cooling towers/evaporative condensers? If so, have they been registered with the local authority?			
Are there any infrequently used showers, and so forth, on the premises? If yes, are they regularly operated to prevent the build-up of scale, and so on, attractive to Legionellae?			
Are you satisfied that your water installation is being properly maintained so as to minimize the risk of a Legionellae problem?			
Noise			
Do your company's processes suggest that noise levels might border upon the action levels? If so, has any measurement been done to confirm this?			
If you operate at any of the three action levels, do you comply with the requirements shown in Table 13.1 of the 'Noise' section of this chapter?			
Do you understand that even if your premises are not at or near the action levels prescribed, the company should still try to reduce noise as far as possible? Note that this duty is now reinforced by the DSE regulations, and that many DSE 'users' have complained that they work in noisy conditions arising from A/C, copiers, printers, and so on.			
Pressure systems			
Does your premises have any pressure systems where the contents are 0.5 bar or above atmospheric pressure, or where there is steam present?			
Have you discussed the regulations on pressure systems and so on with your insurance carrier to ensure that you are complying with the new requirements?			

Reproduced from *What Every Manager Needs to Know About Health and Safety* by Ron Akass, Gower, Aldershot, 1995.

Checkpoints	Action required Yes	No	Action by
Health and safety information for employees			
Do you display sufficient copies of the statutory poster relating to these regulations, completed to show the two addresses required?			
If not, have you issued every employee with the leaflet version of the poster obtainable from HMSO?			
Employers' Liability (Compulsory Insurance) Act			
Do you prominently display a copy of your current certificate for this insurance? You must do so in every workplace.			
Thermometers			
Are there sufficient thermometers readily available to enable staff to check the temperature in their workplaces?			
Safety signs			
Are all the safety signs in your buildings in conformance with these regulations: that is, have all incorrect, *ad hoc*, superseded signs been replaced?			

Reproduced from *What Every Manager Needs to Know About Health and Safety* by Ron Akass, Gower, Aldershot, 1995.

PART THREE

THE SINGLE MARKET

❖

14

EUROPEAN UNION HEALTH AND SAFETY DIRECTIVES

❖

The much heralded birth of the fully-integrated European Union, which, although styled 'Euro-92', did not become a reality until 1 January 1993, has brought with it a bombardment of new health and safety legislation, and the prospect of much more to come.

Many businesses view with considerable anxiety a mass of legislation the 'added value' of which – in health and safety terms – must be questionable, and wonder whether it represents any improvement at all on the excellent legislation we already have, which is not only a model of its kind, but can claim to be the 'mother' of occupational health and safety law. On the one hand, therefore, we have our own primary health and safety legislation – The Health and Safety at Work Act 1974 – which is and will remain our principal occupational health and safety legislative instrument. On the other hand, in addition to compliance with HASAWA and its subordinate regulations, we have also to comply with the EU-driven regulations. For some EU members these problems do not exist; at least one member state ratified the Framework Directive without a single alteration, suggesting that before 1993 they had no health and safety legislation at all – or none worth keeping!

INTERPRETATION

Of the EU health and safety directives ratified here so far, only the Health and Safety (Display Screen Equipment) Regulations 1992 deal with matters not already addressed in varying degrees by existing legislation. Unfortunately much of the content of the new regulations is too vague and imprecise to be of real benefit in advancing the cause of workplace health and safety. Far from acting as a catalyst for improvements, the regulations have succeeded in creating chaos, confusion, anxiety and frustration for many businesses, particularly those at the smaller end of the spectrum.

Nowhere is this more true than in respect to the display screen equipment (DSE) regulations which are covered in detail in Chapter 16. If the intention of these regulations is to bring about some kind of similarity of standards and conditions among those who work with display screens – VDUs, as we are accustomed to calling them – it cannot succeed. It cannot succeed where more than one person shares the same office, and it certainly cannot succeed across the EU!

The DSE regulations deal with a number of matters pertaining to the working environment – hardware, software, furniture and environmental factors – some of which have defied objective assessment for decades; for they simply are not capable of fitting tidily within the framework of legislation. Furthermore it is virtually certain that where two or more people share the same office, there will be differences of opinion as to whether it is too warm or too cold, whether the lighting is too bright or too dim, whether the area is noisy or not, and so on.

When the sun shines in winter, some DSE users will welcome it after so many grey days, although it might cause some glare on screens, however well positioned. Yet even if DSE screens are so arranged that the sunshine creates no glare effect at all, there will still be some staff who want the blinds closed. How can legislation resolve this common environmental and behavioural problem which occurs whenever there is winter sunshine? Perhaps employers should provide every single DSE 'user' with their own exclusive sealed office, offering each individual total control of their environment and workstation, so that there would be no more complaints about temperature, noise and so on . . . until a new cycle of complaints – isolation, loneliness and depression – starts to arise!

In many industries and occupations there are different opinions as to whether or not particular jobs create display screen equipment 'user' status for those that work at them. Some industry groups cannot agree about the user status of people who do the same job, even having the same job title, albeit for different employers (see 'Health and Safety (Display Screen Equipment) Regulations 1992', page 175).

If it is impossible to achieve consensus within one industry group, how can there be a common approach throughout the country, far less the entire European Union?

COMPLIANCE WITH DIRECTIVES

Add to the variances in interpretation described the fact that attitudes among member states towards compliance with EU Directives vary widely, and it becomes clear that far from being a 'Common' or 'Single' market, there is in fact a 'singularly uncommon' market in terms of approach to compliance with these directives.

Far from creating a level playing field for competition, these regulations must have the opposite effect than that intended. At one stage it was hinted that we should take a more 'Latin' attitude to compliance with EU directives, implying something between being 'laid-back' and doing nothing at all!

However frustrating it may be to work to achieve the required standards while others are doing nothing, it would be wrong to do otherwise than press ahead. Any suggestion that we should be 'judgemental' about compliance with law is the beginning of the downward slide. Our traditional philosophy that laws are made to be obeyed must not be compromised.

Meantime, as stated in the introduction to this book, these and many other regulatory instruments are being reviewed to try to sort things out.

STANDARDS

Some regulations which have appeared to date do not provide clear guidance as to the requirements or standards to be achieved. Both quantitative and qualitative standards remain vague. For example, employers being enjoined to comply with requirements that are qualified by terms such as 'suitable' and 'sufficient'.

Given the duty of all employers to provide a healthy and safe workplace and plant, safe systems of work, and so on, it follows that ensuring display screen equipment is safe, that untrained employees do not lift heavy loads, that tools and equipment are in good order, and so forth, are elements of the existing duty. The claim that many parts of the new regulations repeat existing requirements is best exemplified by the Management of Health and Safety at Work Regulations (MHSW), which call for risk assessments and procedures to deal with serious and imminent danger *as well as requiring arrangements to be made for the effective planning, organization, control, monitoring and review of the preventive and protective measures.*

THE HEALTH AND SAFETY POLICY

Employers who conscientiously review their existing health and safety policies, update them when appropriate, and ensure that all employees are kept aware of changes, must by implication include procedures for dealing with serious and imminent danger – a duty which they have had in accordance with Section 2 of HASAWA since 1974. It is difficult to understand how *any* safety policy could be developed which does not address such matters.

ASSESSMENTS

Many of the new regulations require the compilation of an 'assessment' – a term relatively new to us, although it first became part of the health and safety vocabulary with the advent of the Control of Substances Hazardous to Health

Regulations 1989 (COSHH) (see Chapter 10).

Experience has shown that there is widespread misunderstanding and confusion regarding assessments. This should not surprise us. Hitherto employers have been accustomed to being told precisely what the law demands – and what will happen if they do not comply. They are not accustomed to the notion of regulations that spell out the broad objectives, and then leave it to them to apply the principles to their own enterprises. None the less, this approach has much to commend it, and it was fortunate that COSHH gave a foretaste of what was to come.

Each of the six new health and safety regulations is discussed in detail in Chapters 15–20.

MANAGEMENT ACTION CHECKLIST 14

European Union Health and Safety Directives

Checkpoints	Action required Yes	No	Action by
Has your company considered the implications of the six EU-motivated Regulations in so far as its operations are concerned?			
Has a 'steering committee' been established to monitor progress toward compliance with these Regulations?			
Does the company have a copy of each of the Regulations and its associated Code of Practice?			

15

MANAGEMENT OF HEALTH AND SAFETY AT WORK REGULATIONS 1992 (MHSW)

❖

Had it not been for the strength of feeling expressed during the pre-enactment consultative process, we might have been asked to comply with a regulation with the bland and uninspiring title originally proposed – The Health and Safety (General Provisions) Regulations! Fortunately heed has been taken of the need to make legislation a little more user-friendly, at least in the way it is styled.

MHSW is the most important of the six sets of health and safety regulations introduced here to ratify EU directives covering a range of health- and safety-related subjects. All six sets of regulations take effect from January 1993, although there is a transition period for three of the regulations. The workplace (HSW) regulations come into full effect on 1 January 1996, the display screen (DSE) regulations on 31 December 1996, and the work equipment regulations (PUWER) on 1 January 1997.

The importance of MHSW derives from the fact that it is our version of the EU Framework Directive on workplace health and safety, under whose sponsorship the five 'daughter' directives were developed. These five directives are as follows:

1. Provision and Use of Work Equipment Regulations 1992 (PUWER) (see Chapter 20).
2. Workplace (Health, Safety and Welfare) Regulations 1992 (HSW) (see Chapter 17).
3. Health and Safety (Display Screen Equipment) Regulations 1992 (DSE) (see Chapter 16).
4. Personal Protective Equipment at Work Regulations 1992 (PPE) (see Chapter 19).
5. Manual Handling Operations Regulations 1992 (MHO) (see Chapter 18).

Whereas these five subordinate directives deal with specific aspects of health and safety, MHSW spans a range of requirements which relate to the management and organization for health and safety in the workplace. They are the impetus for employers to review all their health and safety arrangements, and to put in place effective arrangements not only to create and maintain a healthy and safe working environment, but to respond quickly and effectively if things go wrong.

In some ways MHSW is analogous to our primary legislation – the Health and Safety at Work Act 1974, which deals in principles, and has been referred to as an 'umbrella' Act under whose auspices regulations covering specific aspects of health and safety have been introduced here. In terms of content, MHSW is far-ranging; it is a compendium of subjects, and in other circumstances might have appeared as four or five separate regulations. Although it has been said that MHSW essentially 'fleshes out' the best in our own health and safety legislation, in particular HASAWA, the fact is that there is *much more detail* in the requirements of MHSW, and unquestionably a lot of additional work to be done to demonstrate compliance.

While some of the new requirements 'add value', these are overshadowed by vague and muddled sections which can do nothing to improve understanding, or to promote the health and safety message generally. A prime example of this is in relation to the duties of employees.

HEALTH AND SAFETY RESPONSIBILITIES OF EMPLOYEES

The duties of employees appear in Sections 7 and 8 of HASAWA. They are clear, succinct and capable of being readily understood by employees:

> Section 7
>
> (a) Look after yourself while at work, and take care for the health and safety of others who might be affected by your acts or omissions.
>
> (b) Co-operate with your employer in all the measures which he takes to discharge his statutory duties relating to health, safety and welfare.
>
> Section 8
>
> Do not intentionally or recklessly interfere with or misuse anything provided in the interests of health, safety and welfare.

These sections can be summarized as follows: while at work look after yourself and others, co-operate and comply with your employer's health and safety arrangements, and do not interfere with equipment provided for safety and so on.

It is clear that these three statements have relevance and meaning across the entire spectrum of human endeavour. They are easy to learn, easy to teach, easy to understand and applicable to every workplace scenario possible. There are no breaches of safety that an employee could commit which are not covered by Sections 7 or 8 of HASAWA.

EU-driven legislation

Compare the above with the employee's duties in Regulation 12 of MHSW, which was drafted to ratify the EU directive:

> (1) Every employee shall use any machinery, equipment, dangerous substance, transport equipment, means of production or safety device provided to him by his employer in accordance both with any training in the use of the equipment concerned which has been received by him and the instructions respecting that use which have been provided to him by the said employer in compliance with the requirements and prohibitions imposed upon that employer by or under the relevant statutory provisions.
>
> (2) Every employee shall inform his employer or any other employee of that employer with specific responsibility for the health and safety of his fellow employees –
>
> (a) of any work situation which a person with the first-mentioned employee's training and instruction would reasonably consider represented a serious and immediate danger to health and safety; and
>
> (b) of any matter which a person with the first-mentioned employee's training and instruction would reasonably consider represented a shortcoming in the employer's protection arrangements for health and safety, insofar as that situation or matter either affects the health and safety of that first-mentioned employee or arises out of or in connection with his own activities at work, and has not previously been reported to his employer or to any other employee of that employer in accordance with this paragraph.

Until the advent of the EU-driven legislation, an employee's responsibilities under Sections 7 and 8 of HASAWA contained 111 simple, well-chosen and meaningful words. All that MHSW does is to add a further 211 words (190 per cent) which produce nothing not already expressed or clearly implied by the duties of employees enshrined in the Health and Safety at Work Act 1974. Looking after oneself – a key element of the employee's duties in accordance with HASAWA – must mean working as instructed, reporting that which is wrong or unsafe, and so on.

What is the point of asking an employee to report concerns provided that they have not 'previously been reported'? How will the reporting employee know, until he or she reports a concern, whether or not it has already been reported, and what is there about these reporting requirements that is not already an integral part of every employer's health and safety arrangements?

Employers have a duty in accordance with HASAWA Section 2(3) to produce a safety policy, and to bring it and all revisions/updates to the attention of all their employees (see Chapter 3).

A safety policy would be ineffective if it did not contain an injunction to employees to report *anything* untoward or dangerous in the workplace. Since a safety policy is required by statute, it follows that if the policy's contents are 'reasonable', failure to comply with them would be a disciplinary matter, and might, in the most serious circumstances, expose the offending employee to official reprobation. The inclusion in a company's safety policy of a requirement for employees to report dangers and prima facie safety breaches is essential. Indeed it would be 'unreasonable', even negligent, for such a requirement to be omitted from a safety policy.

We know that all EU-driven regulations represent a degree of compromise; on the evidence of this regulation, we appear to have overdone compromise to the point of absurdity!

COMPLIANCE WITH MHSW

Many of the regulations in MHSW repeat existing duties, or overlap with them; for example, Regulation 3 relates to risk assessments, but these are already demanded in respect to hazardous substances to comply with the COSHH regulations of 1989. It will not, therefore, be necessary to produce a further COSHH assessment within the overall assessment exercise called for in Regulation 3 of MHSW.

Where old and new legislation contains different words, however, as in the case of employees' duties already described, it is incumbent upon those affected to comply with *both* the HASAWA and MHSW versions. Employers will have to ensure that where one regulation is 'general' in its description of a duty, and 'specific' in another description of the same duty, the more onerous of the two requirements must be complied with.

THE ELEMENTS OF MHSW

Each of the constituent regulations within MHSW are reviewed; where the requirement is new to the United Kingdom this is stated.

REGULATION 3 – RISK ASSESSMENT (NEW)

This regulation calls for a 'risk assessment' to be carried out by every employer and self-employed person, and the results produced in writing or electronically if the enterprise employs five or more people.

The risk assessment should address significant risks inherent in the enterprise which might affect employees or other persons. The objective is to ensure that suitable measures are taken to protect those likely to be exposed to risk, thus complying with relevant health and safety legislation.

Although there is no requirement to duplicate work already done in terms of risk assessment in order to comply with other health and safety legislation – for example, COSHH, MHO, and so on – it makes sense to integrate other assessments into this exercise, not only for the sake of completeness, but also to reduce the volume of separate documentation on health and safety. Unless some rationalization occurs, there will soon be more paper in existence relating to occupational health and safety than for the running of the business itself!

For those businesses that have effectively complied with the duty to produce a health and safety policy, little or no further work is needed to meet the requirements of this regulation. In order to develop an effective company health and safety policy – a statutory duty since 1974 – a risk assessment should already have been carried out. Not to have done so means that the company safety policy has been produced disregarding the hazards inherent in the business.

Where risk assessments have been carried out as a precursor to preparing the company safety policy, all that is needed is to ensure that they are regularly reviewed and tested.

Developing the risk assessment – considerations

Scope: It is not the intention that risk assessments should cover every possible contingency, rather that they take account of risks that could reasonably be expected to arise; the focus should be on significant risks.

Ranking risks: For many firms, the ranking of risks in terms of probability and seriousness will be intuitive. In these cases simple classifications such as 'low', 'medium' and 'high' risk will suffice to determine what measures should be taken to mitigate the effects of accidents and emergencies.

Frequently accident records, whether those of the company concerned, or the industry group to which the company belongs, may be helpful by advising on the frequency and seriousness of risks in that industry.

Other methods of calculating risk include giving numeric values to seriousness and probability, and multiplying the two. This will be helpful in determining where the most effort is needed to reduce the risks – and for the process of initiating procedures to deal with emergencies, which is the subject of Regulation 7 below.

Again industry groups and trade or professional associations might be of assistance in producing risk tables for their members. For businesses with many locations (e.g. retail chains) considerable resource savings can be achieved by developing 'model' assessments which can be used for all locations in which the operations – and risks – are the same. Even though there may be some locations with slightly different operations to others, it could be beneficial to develop a company-wide assessment, leaving local management to delete items that are not relevant to them.

Types of risk

Every business has risks of one kind or another, although in some cases these will be small enough to disregard. Some risks will be peculiar to the business in question, or be a feature of a particular industry or profession.

Sadly there are now a number of workplace risks that are an unpleasant feature of the closing years of this century – violence and terrorism, vandalism and arson. Violence may occur in connection with robbery – often armed robbery. This is becoming commonplace in banking, among building societies, post offices and other areas where the felon believes that money can be obtained. Alcohol and drugs may be causal factors in hospital casualty areas and the licensed victualling trade, whereas violence induced by frustration can occur where civil servants or local government officers interface with the public, for example in the DSS, housing offices and social services.

Terrorism is no longer confined to government buildings, and even where a business is unlikely to become a terrorist target itself, it could be located close to a government office or other possible target. In multi-occupancy buildings, there might be some high-risk tenants; where such buildings also have car parks for use by all tenants, the danger of car bombs is a consideration, and there should be suitable security measures. Regulation 9 below deals with co-operation and co-ordination in multi-occupancy buildings.

Vandalism and arson are not confined to any particular industry. Every employer therefore needs to take account of the possibility of attacks of either sort, where such attacks have the potential to endanger employees or others.

Risks directly related to business operations

In large businesses, the initial identification of risks could be carried out by individual departments or functions, or be summarized elementally (e.g. machinery, transport, electrical). Miscellaneous considerations are as follows:

1. Risks associated with particular worker groups (e.g. night watchmen, counter staff, cleaners, lone workers, new starters).
2. The assessment must be recorded in writing (or stored electronically) if the workforce totals five or more. It should either include the conclusions and preventative and protective measures, or state where these are kept.
3. Principles of risk avoidance:
   (a) avoiding or combating risk at source;
   (b) adapting work to the individual – ergonomics;
   (c) keeping abreast of technical and technological progress;
   (d) giving priority to measures that protect everybody.
4. Assessments properly produced to satisfy other regulations (e.g. COSHH, MHO) do not have to be repeated.

REGULATION 4 – HEALTH AND SAFETY ARRANGEMENTS

The requirements of Regulation 4 cannot be termed 'new' as they are largely a repeat of Section 2(3) of HASAWA – duty to produce a health and safety policy (see Chapter 3).

REGULATION 5 – HEALTH SURVEILLANCE (NEW)

Regulation 5 emphasizes the employer's duty to ensure that his employees are provided with health surveillance appropriate to the risks to their health and safety identified by the risk assessment (see Regulation 3).

Some health surveillance needs may have been identified during the COSHH assessment. The risk assessment exercise under these MHSW Regulations calls for health surveillance in the following circumstances: if there is an identifiable disease or adverse health condition related to the work concerned; if valid techniques are available to detect indications of the disease or condition; if there is a reasonable likelihood of the disease or condition under the particular conditions of work, and if surveillance will improve the protection of the health of the employees to be covered.

Although the main objective of health surveillance is to provide early warning, it also serves as a check on the effectiveness of control measures, provides feedback on the accuracy of the risk assessment, and highlights any individuals at increased risk.

REGULATION 6 – HEALTH AND SAFETY ASSISTANCE (NEW)

Until these regulations were enacted there was no statutory duty to appoint safety officers (or advisers as they are now referred to), except for some aspects of construction.

The regulations now call for the appointment of one or more 'competent' persons to assist the employer in the discharge of his statutory health and safety duties. The persons(s) appointed may be employees or consultants, the important criterion being that of competence.

There is emphasis on co-operation between those appointed, and they must of course be fully conversant with *all* the arrangements and procedures in place or in plan where these relate to health, safety and welfare. This includes receiving details of all temporary employees.

These appointments in no sense absolve employers from responsibility for compliance with all relevant health and safety legislation.

Competence

The definition of 'competence' remains, as it always has been, an abstract one, and it is the employer's job to decide whether those he appoints are competent. The guidance describes 'competence' as of having *sufficient training and experience or knowledge and other qualities* to enable a person to properly assist the employer in discharging his health and safety responsibilities.

In deciding whether a proposed safety adviser is competent, the employer should consider the following:

1. Knowledge and understanding of: the work involved; the principles of risk assessment and prevention; and the current health and safety 'state of the art'.
2. Capacity to: evaluate situations that might arise in the enterprise; to design solutions; to communicate effectively at all levels; and generally to promote the aims and objectives of workplace health and safety throughout the business.

There will be occasions where a particular problem is beyond the skill of the appointed adviser. Given the present fast-moving and dynamic nature of occupational health and safety, these occasions are likely to occur more frequently. This should not reflect upon the appointed adviser/s, who should be mature enough to recognize their limitations. In such cases, specialist advice should be sought.

At present the spectrum of safety provision ranges from safety departments staffed with a number of professionals in health- and safety-related disciplines through to firms having no health and safety expertise at all! Between these extremes may be offices in which the administration manager finds himself burdened with health and safety responsibility in addition to the myriad other duties that the position embraces. Fortunately for their employers, the administration or office managers are sensible enough to demand some training to fit them for the safety responsibility, and if they are lucky, they will be sent away on a course or two in order to gain the necessary knowledge. The less fortunate will be told that they can attend a one-day seminar on general safety – or worse, a day devoted to only one facet of the safety agenda. When delegates are otherwise untutored in health and safety, one-day seminars are comparable to the Victorian 'crammer', except that the unfortunate administration manager has not even had the basic schooling in the first place!

This regulation should serve to prevent untrained or partially trained staff being appointed to positions with a safety advisory responsibility.

REGULATION 7 – PROCEDURES FOR SERIOUS AND IMMINENT DANGER (NEW IN PART)

Although this regulation is not entirely new, it does call for a reappraisal of an employer's existing emergency arrangements, which must of course take account of the conclusions of the risk assessment exercise carried out to comply with Regulation 3 of MHSW. Regulation 7 is concerned with the following:

1. The establishment of proper emergency procedures, with adequate staffing to implement those procedures in so far as emergency evacuation is concerned. This is in effect a restatement of the requirements of a fire certificate, with the addition of procedures to cover evacuation for reasons other than fire (e.g. a bomb threat, danger of explosion, other major plant failure).
2. The nomination/appointment of sufficient competent staff to implement the evacuation procedure. There is nothing new in this, but the emphasis is on staffing emergency evacuation, nothing else. There is, and remains, a problem with emergency roles. For example, where a building is evacuated as a result of a bomb threat, the police prefer employee volunteers to accompany their search parties as guides and to speed up the search process generally. This is not a statutory duty, and even where volunteers are forthcoming, it should not be carried out without prior consultation with the employer's insurance carrier.
3. Preventing employees from entering any area where there are health and safety risks until they have been properly trained and given instruction on the hazards present.
4. Where practicable, inform persons at work exposed to serious and imminent danger of the hazard and the steps being taken to protect them from it.
5. Enable persons referred to in item 4 to stop work immediately and proceed to a place of safety if they are exposed to serious, imminent and unavoidable danger, and unless there are documented procedures

justifying exceptions, prevent such persons from resuming work where serious and imminent danger remains.

The regulation deals with what we properly understand to be emergency procedures, and instead of the lengthy user-unfriendly title chosen for it, readers could have at once understood and been more comfortable with 'Emergency procedures'.

Although most businesses have recognized the need for some organization to deal with emergencies, there are some who do nothing until they have a serious incident – often at the expense of safety as well as the commercial viability of the business. For those who have already developed emergency procedures, all that is needed is a review to ensure that the arrangements made are still valid and cover the risks identified to comply with Regulation 3 of MHSW.

In many cases, particularly in the commercial sector, serious dangers might be confined to lift entrapment, fire and terrorism. In the industrial sector there may be many areas of concern (e.g. the danger of explosion, release of toxic substances).

Emergency procedures

A common failing is to highlight the roles of 'key players' at the expense of guidance to others. One of the first considerations is to remove from areas of danger those who have no part to play.

Regulation 7 refers to danger areas: that is to say, areas where the level of risk is such that employees should not enter without special precautions being taken. In addition to static areas of this kind, there will be others of a transitory nature such as maintenance and renovation work, engineering changes to manufacturing areas, and so on.

Item 5 in the list above refers to the right of workers to vacate an area if they perceive danger, and to remain in a secure place until it is safe to return. The emphasis here is on a worker's right to use his own initiative about moving to a place of safety in the event that supervisors are not present to give approval to do so. This is a curious clause, suggesting that there are employers who would penalize employees for saving themselves!

The spirit of Regulation 7 is that of communication – making sure that clear procedures exist to cover the emergencies that could arise as highlighted by the risk assessment, and that these procedures are properly communicated. This communication should be repeated as appropriate, and should always be conveyed to any new employees who will be affected as part of their induction training (see Regulation 11). Safety advisers appointed in accordance with Regulation 6 should be involved in the development of emergency plans and procedures.

REGULATION 8 – INFORMATION FOR EMPLOYEES

Regulation 8 requires clear and comprehensible information to be provided to all employees, including trainees and those on fixed-duration contracts, on all the health and safety risks and mitigating measures adopted by their employer.

Care should be taken to ensure that the information given is understandable, taking account of the level of training, knowledge and experience of those involved. Special arrangements need to be made where employees do not readily understand English, in which case there might also be a need for symbols or diagrams to supplement notices written in English.

The important duty is to ensure that information is 'comprehensible'.

REGULATION 9 – CO-OPERATION AND CO-ORDINATION (NEW)

The objective here is to ensure that where two or more employers share a workplace, either permanently or on a temporary basis, there is co-operation and co-ordination on all health- and safety-related matters. These will include emergency evacuation, and the disclosure by each employer to the other of any aspects of their operations which could give rise to danger.

Self-employed persons working as contractors for an employer, or working on the premises of another self-employed person, are to be regarded as sharing the workplace for the purposes of this regulation.

Where there are shared premises, but one of those sharing is the controller of the worksite, other employers or self-employed persons working there should co-operate in the measures taken, and provide appropriate health and safety information to the site-controlling employer. This requirement also extends to co-operation with landlords or managing agents where they have responsibility for or control of the premises.

Controlling employers should ensure that new minor employers or self-employed persons understand the arrangements made, and can integrate themselves into them.

None of the contents of this regulation are really new, the duties being expressed in more general terms in existing legislation. However, it is fair to say that in a great many 'shared' situations there is no contact between the parties involved at all! As a consequence, health and safety matters are neglected, and the dangers to occupants in the event of emergency are greater than in single-occupant business premises.

This regulation is therefore timely in that it requires those who share premises to co-operate in terms of general safety, as well as in respect of emergency evacuation. It is particularly valuable in dealing with the difficulty experienced in business premises occupied by a number of companies, where an acceptable date for carrying out emergency evacuation drill cannot be agreed. A consequence of this difficulty is either that the more responsible tenants are concerned that the landlord has not arranged drills at the appropriate frequency, or that landlords become frustrated by the unwillingness of some tenants to agree a date for evacuation drills.

Where there is no controlling employer in a shared premises, it may be necessary for the occupants to agree to appoint a health and safety co-ordinator from among themselves, in order to ensure compliance with applicable statutory requirements.

REGULATION 10 – PERSONS WORKING IN HOST EMPLOYERS' OR SELF-EMPLOYED PERSONS' UNDERTAKINGS (NEW)

It is the duty of employers and the self-employed to provide the employer of any person who may come to work in their premises with complete and comprehensible information about any risks that may exist in the premises, and of the measures that they have taken to deal with these risks, including the names of those responsible for the implementation of evacuation procedures.

The risks referred to here are those identified by the risk assessment exercise carried out to comply with Regulation 3 of MHSW.

Again, though not entirely new, this is another requirement of MHSW which does give a different slant to existing law and practice. It has been recognized for many years that when awarding contracts, the client employer should emphasize his safety policy and procedures, and require prospective contractors to do the same. The recommendation has been to include a review of the prospective contractor's health and safety policy, accident record and other related matters in the case of construction contracts or other work with a potential for accidents. Contractors were also advised to request a copy of the client employer's safety policy, and it is difficult to envisage how a tendering contractor could properly develop his tender without knowing the health and safety requirements of the prospective client.

As such, the regulation is not concerned with the dangers which a contractor might introduce into a client employer's premises; these are addressed by Section 3 of HASAWA. It is concerned solely with the duty of an employer to the workers of his contractors when they are working on his premises.

REGULATION 11 – CAPABILITIES AND TRAINING

Here is yet another permutation of existing law, and it is difficult to understand why the perfectly adequate and comprehensive requirements of Section 2(2)(c) of HASAWA required further amplification.

Section 2(2)(c) of HASAWA requires an employer to provide sufficient information, instruction, training and supervision as is necessary to ensure, so far as is reasonably practicable, the health and safety at work of his employees. The guidance on this requirement covers matters such as induction and repeat training, appropriate rules and procedures, and so on.

Capabilities

Employers are required to take account of the capabilities of their employees as regards health and safety when entrusting tasks to them. In short, do not ask them to do things which they are not capable of!

In respect to training, the regulations repeat the guidance to the 1974 Act, namely that training shall be provided as follows:

1. On being recruited (induction training).
2. On being exposed to increased risks as a result of:
   (a) being transferred or given a change of responsibility;

(b) introduction of new or changed equipment;
(c) new technology;
(d) new or changed systems of work.

Qualifications

Training must be repeated periodically where appropriate, adapted to take account of changes in respect to the health and safety of those concerned, and take place during working hours. Where it is not possible to complete training during normal working hours, the employer's rules for extended time at work must be applied.

The only novel element of this regulation is that related to capabilities, and it is difficult to understand the motive for stating what is patently obvious, not only from a health and safety standpoint, but measured against the entire spectrum of work activity.

Surely the basic task of management is to ensure that those whom they direct are capable of carrying out the tasks assigned to them. Anything short of this would be commercial suicide. QED! If we are to make safety integral and equal to everything else in the workplace, why separate it in this way?

REGULATION 12 – EMPLOYEES' DUTIES

These duties have been discussed in detail in the opening paragraphs of this chapter.

REGULATION 13 – TEMPORARY WORKERS (NEW)

This regulation is complementary to Regulation 10 which dealt with the provision of health and safety information to employers whose employees will be working on the client employer's premises. In this case, the concern is to ensure that temporary workers, whether engaged directly by the employer or through an agency, also receive all the necessary information about the health and safety arrangements, the risks associated with the work, and the measures taken to ameliorate them, in the premises where they will be working.

Documentary evidence of the provision of such information is important, and temporary staff agencies will in turn need to maintain evidence that they have passed on information to the staff they select for the work in question. Specifically, the requirement is to provide information as follows:

1. To temporary staff engaged direct by the employer:
   (a) any special occupational qualifications or skills required by the temporary employee in order to carry out the work safely; and
   (b) any health surveillance required to be provided for that employee under statutory provisions (e.g. COSHH, MHSW – Regulation 5).

   The information is to be provided by the employer before the employee commences the temporary work.
2. By the prospective employer or self-employed person to the temporary

staff to be provided by an agency: (as in 1(a) and (b) above).

3. By the prospective employer or self-employed person to the agency providing temporary staff:
   (a) as in 1(a) above; and
   (b) the specific features of the jobs to be filled by those employees, in so far as these might affect their health and safety.

CONCLUSION

Now that the full weight of these regulations and the other five sets comprising the first clutch of EU-generated health and safety regulations is beginning to be felt, the establishment line is that there really is not as much work to be done as at first appears. It is for the reader to judge the truth of this assertion – perhaps after reading the list of points that follow in the management action checklist.

MANAGEMENT ACTION CHECKLIST 15

Management of Health and Safety at Work Regulations 1992 (MHSW)

Checkpoints	Action required Yes	No	Action by
Is a copy of MHSW and its ACOP held?			
Is a working party/steering committee needed to ensure MHSW is properly implemented?			
Risk assessment. What action has been taken to develop and publish an assessment as demanded by MHSW?			
Regulation 4 – health and safety arrangements (see Chapter 3) – safety policy management review. If health surveillance is required following risk assessment, what are the arrangements?			
Health and safety assistance (see Chapter 2) – organization for health and safety management review. Apart from fire emergency procedures, what others have been identified as being necessary following the risk assessment? Are they ready?			
Information for employees – Regulation 8. Is the company communications package on health and safety generally – and, in particular, on MHSW – 'comprehensible' to employees?			
If the company share any premises, has consideration been given to the requirements of Regulation 9?			
What communication on health and safety is given to contractor managements and their staff, and to other visitors to company premises?			
Regulation 11 – capabilities and training. Have company health and safety training arrangements been reviewed, particularly as regards induction and refresher training? Even if arrangements were satisfactory, a substantial training effort is needed to acquaint all employees with their new/additional statutory responsibilities in accordance with Regulation 12.			

Reproduced from *What Every Manager Needs to Know About Health and Safety* by Ron Akass, Gower, Aldershot, 1995.

Checkpoints	Action required Yes	No	Action by
What procedures now exist to ensure compliance with the duty to provide full health and safety briefing to temporaries, whether recruited direct or through agencies?			

16

HEALTH AND SAFETY (DISPLAY SCREEN EQUIPMENT) REGULATIONS 1992 (DSE)

A serious review of all six health and safety regulations enacted with effect from 1 January 1991 to ratify EU directives would reveal that only one of them – the DSE regulations – is entirely new. The other five are replays or nuances of existing health and safety law and regulation. The description 'display screen equipment' is not a term we have been accustomed to use, and hopefully we shall continue to call the equipment 'visual display units' (VDUs).

These regulations attempt to bring objectivity to what is and will remain a highly subjective area of work. Once the subject of VDUs becomes linked to environmental matters, such as temperature, light, glare, noise and space – however logical this is in principle – the outcome must be totally subjective, and no bureaucracy anywhere can change that fact!

In any enclosed or open office or workplace where more than two people are working in proximity, there *must,* by definition, be different views about the environmental factors – indeed, about the furniture and equipment as well. If the EU directive and our ratifying regulations are intended to bring about some uniformity in standards of working conditions for DSE/VDU users across the Union, they will fail utterly. They will not even bring about consistency in a single office in the UK if there are more than two users working in it. The best that these regulations can bring about is a heightened awareness of ergonomics and the health risks of poor posture, coupled with the elimination of the *extremes* of poor working conditions and equipment; but *only* the extremes.

For the great majority of workplaces where VDUs are in use, the effort to resolve the various issues will be huge – and out of all proportion to the benefits that will derive. Fortunately, these regulations, although already in force, do not have to be complied with fully until 1 January 1997. This will allow the following to happen:

1. The government review of all EU-driven legislation currently in force, aimed at simplifying or removing unnecessary and burdensome

legislation, will be published.

2. The EU will eventually determine precisely what standards of compliance it requires for DSE. At present there is no EU standard, incredible as this may seem, which means that at present we have to comply with a directive which cannot explain precisely what it wants.

When we eventually receive this information, it should not entail too much reworking for those firms conscientious enough to have already started upon their assessments, as called for in the regulations.

The guidance to the regulations currently calls for some work in relation to the frequency of update/review of an assessment once it has been made. Seven actual reasons are listed for why an assessment, once concluded, should be reviewed, and there are by implication at least three further reasons.

WHEN TO REVIEW A VDU ASSESSMENT

A VDU assessment or that part of it which is affected should be reviewed in the light of changed circumstances such as changes to the VDU worker population or the capability of individuals, and where other noticeable changes in the workstation have occurred. These include, but are not limited to, the following:

1. Significant changes to the software used.
2. A big change to the hardware (e.g. screen, keyboard, input devices) in use.
3. Basic changes in the workstation furniture.
4. A substantial increase in the time to be spent using the DSE/VDU.
5. A fundamental change in the task requirements (e.g. more speed or accuracy).
6. Relocation of the workstation.
7. If the lighting is significantly modified.

It will also be necessary to review completed assessments if research findings indicate significant new risks, or show that a recognized hazard should be re-evaluated.

Experience suggests that in a dynamic company of, say, about 200 VDU users, there will be a need for one full-time employee to be engaged solely in the process of updating, reviewing and actioning the results of a continuous flow of reassessments.

COSTS

Apart from the ongoing cost of the reassessment activity described, the Health and Safety Executive (HSE) forecast that the cost of implementation of the DSE regulations will average out at £40 for every workstation in the UK. The HSE further intimated that this was a cost that would quickly be recouped from the downturn in sickness absenteeism as a result of the training and improved working conditions that would come about as employers complied with these regulations.

In the author's experience there has been no marked change in sickness absenteeism since the advent of the VDU; in fact it is marginally improved on pre-VDU days. Moreover, where there is a pattern of long-term sickness, the numbers involved are a tiny proportion of the total workforce, *and* it would be wrong to blame even this upon the VDU. Indeed, those who have worked with VDUs and are now unfortunately long-term sick would not, for the most part, attribute their condition to working with VDUs, although a few would include this as just one of a number of contributory factors.

What can be stated about the sickness situation with a great deal more certainty is that in those cases where an employee is determined to obtain some compensation for a real or alleged condition brought about by working as a VDU user, their employer, whatever measures he has taken to meet his obligations under these regulations, will still find himself very much on the defensive.

THE REAL WORLD

The optimum position for a keyboard and screen is directly in front of the user. However, a user might be someone who only uses the equipment for 25 per cent of the working day. It is illogical to position a keyboard and screen permanently in a position that does more to obstruct the user in the performance of their overall job than to assist them. Of course, this can be remedied by extension arms and other aids, but these are not cheap, and this begs the question whether such an expense can be justified on financial or health grounds.

Some people adopt curious postures when seated. A sizeable minority opt to sit down on their crossed legs – posturally a bad position. One of the duties of an employer in general health and safety terms, as well as in respect to these regulations, is to provide sufficient information, instruction, training and supervision to ensure the health and safety of his employees while at work. Thus the dangers of bad posture should be explained. Supposing, though, that the employee who sits in this way either cannot or will not change the habit? Will the employer have to commence disciplinary proceedings to coerce the employee to sit properly – and can he dismiss the employee who persists in the habit?

How can there be sufficient supervision to detect bad posture habits anyway? Supposing an employee does not sit in this fashion, but is generally slovenly about their posture? Presumably the employer will have to require offending employees to sign an affirmation that they have received proper training, understand that their postural habits might cause injury, but wish to continue sitting in the manner that they do. Will this be acceptable? It would not be for a misdemeanour such as not wearing prescribed personal protective clothing.

Perhaps the best clarification comes not in this regulation, but in the guidance to the employee's duties regulation of the Management of Health and Safety at Work Regulations (MHSW) (see Chapter 15). This states, in paragraph 76:

> Employees have a duty under Section 7 of HASAWA 1974 to take reasonable care for their own health and safety and that of others who may be affected by their acts or omissions at work. Towards this end, employees should use correctly all work items provided by their employer, in accordance with their training and the instructions they receive to enable them to use the items safely.

Therefore, strict interpretation of this guidance means that employees must comply with the guidance they receive about posture, and will be in breach of law if they do not. Time will tell whether this contention, even though contained in approved (and therefore enforceable) guidance, will be the view of industrial tribunals or the courts.

The windows are fitted with blinds. They are universally welcomed in summer, but in winter are the cause of much contention. On the few occasions when the sun shines, some of the office occupants welcome the brightness and warmth of the sun; they do not want the blinds down. Others do not like the glare of the sun, and do!

Even allowing for the fact that any new regulation takes time to become fully understood and effectively implemented, the auguries for the DSE regulations seem particularly bleak. What will be necessary here is a pragmatic approach, both to the implementation and the enforcement of the regulations – but the word 'pragmatic' is singularly absent from the guidance to these regulations.

WHAT THE REGULATIONS DEMAND

The following summarizes an employer's duties:

1. To explain the regulations and risks from using DSE/VDUs to all employees, even those who, although occasional users of DSE, are not 'users' within the meaning of these regulations.
2. To determine who the 'users' are.
3. To allow any user who requests it, an eye or eyesight test at the employer's expense.
4. If any user taking an eye or eyesight test receives a prescription for 'special corrective appliances' (glasses or contact lenses), to pay for these to the basic standard.
5. To assess the workstation equipment – VDU/keyboard, furniture, software and environment – against generally accepted criteria (not precise standards – the EU have not published these yet!).
6. Where the assessment highlights a need to change/improve things, to arrange to have the necessary improvements carried out by 31 December 1996, unless delay in doing so would put users or operators at risk.

The curious wording of item 6 suggests on the one hand a transitional period, while on the other requiring the employer to put in hand the improvements immediately if there are 'risks' to employees. If any improvements are needed at all, the presumption must be that this is because employees would be at risk if they were not. Therefore the employer has (a) to decide what must be done by 31 December 1996, and (b) to determine which of the requirements must be put in hand at once if he is not to risk flouting the regulations.

Employee training should include all the matters so far discussed. In addition it should cover rest breaks – where the advice can be summed up as 'little and often'. This does *not* mean stopping work altogether: rather the employee should do some other work for a short period that does not require the same level of concentration and similar hand movements.

ASSESSMENTS

All aspects of health and safety at work should be the subject of dialogue with employees. Nowhere is this more important than in connection with this set of regulations. As already noted, on this subject everyone of those involved will each have their particular views about almost every aspect. To repeat, it is a highly subjective area. Consequently it cannot be a matter upon which management can pronounce solutions in isolation of the views of the employees involved.

All employees who use DSE/VDUs should therefore complete a questionnaire. If the opening questions relate to the time they work with VDU equipment, this will assist in separating the 'operators' (that is, those who use the equipment for periods of roughly less than 2 hours a day) from the 'users' (that is, more than 2 hours a day). In so far as the right to eye tests are concerned, it is only users who have this right – operators do not.

To add to the confusion, there is no precise definition between users and operators that covers every situation, so that while it is easy to identify with some certainty who most of the users and operators are, there is a large rump of borderline cases. This is exemplified by the guidance to the regulations, which lists in the table of examples, nine job descriptions of users, three of operators and four who might be one or the other.

The questions put to employees in the questionnaire must cover the most important elements – the equipment, software, furniture and environment. There is no official version of a questionnaire, but the questions can easily be derived by examining the requirements of the regulations which appear in Appendix 1 to this chapter, which lists the requirements under the following headings:

1. *Equipment* general comment; display screen; keyboard; work desk or work surface; and work chair.
2. *Environment* space requirements; lighting; reflections and glare; noise; heat; radiation; and humidity.
3. *Interface elements* computer and operator/user.

Staff completing the assessment questionnaires have to understand the parameters for each item, and also that the questionnaire affords the respondent an opportunity to amplify any of their answers or to make any additional points. Sufficient space should be available on the form for this purpose, and the form must have provision for the name, department and location (e.g. floor) of the respondent.

Supervision of the assessment and its final compilation and presentation must be entrusted to someone competent to perform these functions. This does not necessarily mean using a consultant; there might be staff who are experienced in DSE operations, or occupational health nurses who could do the work. The important attributes are: knowledge of DSE/VDU operations; an understanding of ergonomics and the importance of posture, especially in relation to DSE work; ability to discuss the completed questionnaires with each respondent; ability to summarize the conclusions and to identify the necessary remedial action; and the ability to present the findings to management.

As this subject is relatively new, especially in so far as assessments are concerned, it is unlikely that anyone could have a grasp of all the issues, which will of course include matters such as computer and furniture design, and the availability of products including adjustable supporting arms for screens, document holders and footrests. Whoever is assigned the task of preparing the assessment, therefore, must feel able to approach others for help with matters where he or she has limited knowledge or expertise.

THE PROBLEMS

The problems in implementing this regulation are too numerous to describe in detail, and the following is a selection in addition to the many problems already described:

- Users have screens and keyboards at an angle, not in front of them.
- User's chair does not meet the adjustability requirements, but none the less they like the chair.
- Space beneath the desk filled with files and work in process.
- General space shortage causes desks to be cluttered.
- Users have particular habits which they are unwilling to change.
- Users who know what suits their workplace best.
- Users who have acquired and use footrests which are unnecessary given their height.
- Purchase of office furniture using aesthetics as the criteria, not ergonomics.

CONCLUSION

When making final assessments, those involved will have no choice other than to make pragmatic judgements, and to take decisions which appear to be out of line with the guidance. In many cases they will find that even though some computers and furniture do not appear to meet the relevant criteria, manufacturers will endeavour to persuade them that it really is acceptable. In other cases manufacturers will not be prepared to give any answer at all.

For employers it is not simply a question of carrying out the initial assessment; it is an ongoing review arrangement to reflect all the changes which are inherent in a dynamic business.

There will be a continuing need to train employees – either new starters, or those who move from operator to user status. Training departments must understand that they have a duty to include DSE/update training in their courses, or to integrate it into some existing offering. They have to address this matter; it will not go away.

Purchasing departments – and executives empowered to purchase furniture and other equipment – must obtain a written certification from potential suppliers that their products comply 100 per cent with the regulations.

Those responsible for planning moves and rearrangements cannot begin to do their work unless they have been fully briefed on these regulations. Executives responsible for reviewing *any* proposals for office moves or alterations should pre-empt the presentation with the question 'Do you fully understand the requirements of the DSE regulations in general, and the environment section in particular?' Unless they can be assured that those involved do have this knowledge, there should be no further discussion on the proposals. Any other course not only exposes the company to sanctions, but will inevitably result in substantial unplanned expense in achieving compliance with these regulations retrospectively.

FURTHER READING

Health and Safety Executive, *Display Screen Equipment Work: Guidance on regulations – Health and Safety (Display Screen Equipment) Regulations 1992*, London: HSE Books. Priced at £5.00. This guidance is essential reading for assessors.

Health and Safety Executive, *Working with VDUs*, no. IND(G)36(L), London: HSE Books. Free from HSE Information Centre. HSE intend to issue further guidance on this subject in due course.

APPENDIX 1

MINIMUM REQUIREMENTS FOR WORKSTATIONS[1]

THE SCHEDULE

EXTENT TO WHICH EMPLOYERS MUST ENSURE THAT WORKSTATIONS MEET THE REQUIREMENTS LAID DOWN IN THIS SCHEDULE

1. An employer shall ensure that a workstation meets the requirements laid down in this Schedule to the extent that:
   (a) those requirements relate to a component which is present in the workstation concerned;
   (b) those requirements have effect with a view to securing the health, safety and welfare of persons at work; and
   (c) the inherent characteristics of a given task make compliance with those requirements appropriate as respects the workstation concerned.

EQUIPMENT

2 (a) General comment

The use as such of the equipment must not be a source of risk for operators or users.

(b) Display screen

The characters on the screen shall be well defined and clearly formed, of adequate size and with adequate spacing between the characters and lines.

The image on the screen should be stable, with no flickering or other forms of instability.

The brightness and the contrast between the characters and the background shall be easily adjustable by the operator or user, and also be easily adjustable to ambient conditions.

The screen must swivel and tilt easily and freely to suit the needs of the operator or user.

It shall be possible to use a separate base for the screen or an adjustable table.

The screen shall be free of reflective glare and reflections liable to cause discomfort to the operator or user.

(c) Keyboard

The keyboard shall be tiltable and separate from the screen so as to allow the operator or user to find a comfortable working position avoiding fatigue in the

---

[1]As contained in EC Directive 90/270/EEC on the minimum safety and health requirements for work with display screen equipment.

arms or hands.

The space in front of the keyboard shall be sufficient to provide support for the hands and arms of the operator or user.

The keyboard shall have a matt surface to avoid reflective glare.

The arrangement of the keyboard and the characteristics of the keys shall be such as to facilitate the use of the keyboard.

The symbols on the keys shall be adequately contrasted and legible from the design working position.

(d) Work desk or work surface

The work desk or work surface shall have a sufficiently large, low-reflectance surface and allow a flexible arrangement of the screen, keyboard, documents and related equipment.

The document holder shall be stable and adjustable and shall be positioned so as to minimize the need for uncomfortable head and eye movements.

There shall be adequate space for operators or users to find a comfortable position.

(e) Work chair

The work chair shall be stable and allow the operator or user easy freedom of movement and a comfortable position.

The seat shall be adjustable in height.

The seat back shall be adjustable in both height and tilt.

A footrest shall be made available to any operator or user who wishes one.

ENVIRONMENT

3 (a) Space requirements

The workstation shall be dimensioned and designed so as to provide sufficient space for the operator or user to change position and vary movements.

(b) Lighting

Any room lighting or task lighting provided shall ensure satisfactory lighting conditions and an appropriate contrast between the screen and the background environment, taking into account the type of work and the vision requirements of the operator or user.

Possible disturbing glare and reflections on the screen or other equipment shall be prevented by co-ordinating workplace and workstation layout with the positioning and technical characteristics of the artificial light sources.

(c) Reflections and glare

Workstations shall be so designed that sources of light, such as windows and other openings, transparent or translucid walls, and brightly coloured fixtures or walls cause no direct glare and no distracting reflections on the screen.

Windows shall be fitted with a suitable system of adjustable covering to attenuate the daylight that falls on the workstation.

Figure 16.A1 Subjects dealt with in the schedule of workstation requirements

1. Adequate lighting
2. Adequate contrast, no glare or distracting reflections
3. Distracting noise minimized
4. Leg room and clearances to allow postural changes
5. Window covering
6. Software: appropriate to task, adapted to user, provides feedback on system status, no undisclosed monitoring
7. Screen: stable image, adjustable, readable, glare/reflection free
8. Keyboard: usable, adjustable, detachable, legible
9. Work surface: allow flexible arrangements, spacious, glare free
10. Work chair: adjustable
11. Footrest

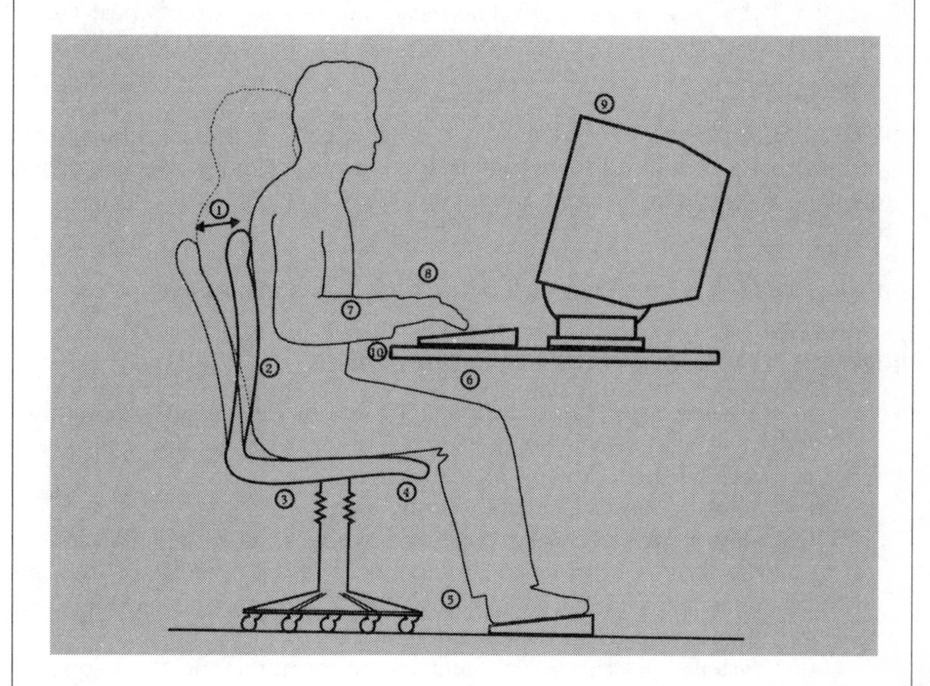

Figure 16.A2 Seating and posture for typical office tasks

1 Seat back adjustability
2 Good lumbar support
3 Seat height adjustability
4 No excess pressure on underside of thighs and backs of knees
5 Foot support if needed
6 Space for postural change, no obstacles under desk
7 Forearms approximately horizontal
8 Minimal extension, flexion or deviation of wrists
9 Screen height and angle should allow comfortable head position
10 Space in front of keyboard to support hands/wrists during pauses in keying

(d) Noise

Noise emitted by equipment belonging to any workstation shall be taken into account when a workstation is being equipped, with a view in particular to ensuring that attention is not distracted and speech is not disturbed.

(e) Heat

Equipment belonging to any workstation shall not produce excess heat which could cause discomfort to operators or users.

(f) Radiation

All radiation with the exception of the visible part of the electromagnetic spectrum shall be reduced to negligible levels from the point of view of the protection of operators' or users' health and safety.

(g) Humidity

An adequate level of humidity shall be established and maintained.

INTERFACE BETWEEN COMPUTER AND OPERATOR/USER

4 In designing, selecting, commissioning and modifying software, and in designing tasks using display screen equipment, the employer shall take into account the following principles:

(a) software must be suitable for the task;
(b) software must be easy to use and, where appropriate, adaptable to the level of knowledge or experience of the operator or user; no quantitative or qualitative checking facility may be used without the knowledge of the operators or users;
(c) systems must provide feedback to operators or users on the performance of those systems;
(d) systems must display information in a format and at a pace which are adapted to operators or users;
(e) the principles of software ergonomics must be applied, in particular to human data processing.

ANNEX A GUIDANCE ON WORKSTATION MINIMUM REQUIREMENTS

1 The schedule to the Regulations sets out minimum requirements for workstations, applicable mainly to typical office workstations. These requirements are applicable only in so far as the components referred to are present at the workstation concerned, the requirements are not precluded by the inherent requirements of the task, and the requirements relate to worker health, safety and welfare.

2 The requirements of the schedule are in most cases self-explanatory but particular points to note are covered below.

GENERAL APPROACH: USE OF STANDARDS

3 Ergonomic requirements for the use of visual display units in office tasks are contained in BS 7179. There is no requirement in the Display Screen Regulations to comply with this or any other standard. Other approaches to meeting the minimum requirements in the Regulations are possible, and may have to be adopted if special requirements of the task or needs of the user preclude the use of equipment made to relevant standards. However, employers may find standards helpful as workstations satisfying BS 7179, or forthcoming international standards (see below), would meet and in most cases go beyond the minimum requirements in the schedule to the Regulations.
4 BS 7179 is a six-part interim standard covering the ergonomics of design and use of visual display terminals in offices; it is concerned with the efficient use of VDUs as well as with user health, safety and comfort. BS 7179 has been issued by the British Standards Institution in recognition of industry's immediate need for guidance and is intended for the managers and supervisors of VDU users as well as for equipment manufacturers. While originally confined to office VDU tasks, many of the general ergonomic recommendations in BS 7179 will be relevant to some non-office situations.
5 International standards are in preparation that will cover the same subject in an expanded form. BS 7179 will be withdrawn when the European standards organization CEN (Comité Européen de Normalisation) issues its multipart standard (EN 29241) concerned with the ergonomics of design and use of visual display terminals for office tasks. This CEN Standard will in turn be based on an ISO Standard (ISO 9241) that is currently being developed. The eventual ISO and CEN standards will cover screen and keyboard design and evaluation, workstation design and environmental requirements, non-keyboard input devices and ergonomic requirements for software design and usability. While the CEN standard is not formally linked to the Display Screen Equipment directive, one of its aims is to establish appropriate levels of user health and safety and comfort. Technical data in the various parts of the CEN standard (and currently BS 7179) may therefore help employers to meet the requirements laid down in the schedule to the Regulations.
6 There are other standards that deal with requirements for furniture, some of which are cross-referenced by BS 7179. These include BS 3044, which is a guide to ergonomic principles in the design and selection of office furniture generally. There is also now a separate standardization initiative within CEN concerned with the performance requirements for office furniture, including dimensioning appropriate for European user populations. Details of relevant British, European and international standards can be obtained from the Department of Trade and Industry.
7 Other more detailed and stringent standards are relevant to certain specialized applications of display screens, especially those where the health or safety of persons other than the screen user may be affected. Some examples in particular subject areas are:

(a) Process control

A large number of British and international standards are or will be relevant to the design of display screen interfaces for use in process control – such as the draft Standard ISO 11064 on the general ergonomic design of control rooms.

(b) Applications with machinery safety implications

Draft Standard pr EN 614 pt 1 – Ergonomic design principles in safety of machinery.

(c) Safety of programmable electronic systems

Draft document IEC 65A (Secretariat) 122 Draft: Functional safety of electrical/electronic programmable systems.

Applications such as these are outside the scope of these guidance notes. Anyone involved in the design of such display screen interfaces and others where there may be safety considerations for non-users should seek appropriate specialist advice. Many relevant standards are listed in the DTI publication *Directory of HCI Standards.*

EQUIPMENT

Display screen

8 Choice of display screen should be considered in relation to other elements of the work system, such as the type and amount of information required for the task, and environmental factors. A satisfactory display can be achieved by custom design for a specific task or environment, or by appropriate adjustments to adapt the display to suit changing requirements or environmental conditions.

Display stability

9 Individual perceptions of screen flicker vary and a screen which is flicker-free to 90 per cent of users should be regarded as satisfying the minimum requirement. (It is not technically feasible to eliminate flicker for all users.) A change to a different display can resolve individual problems with flicker. Persistent display instabilities – flicker, jump, jitter or swim – may indicate basic design problems and assistance should be sought from suppliers.

Brightness and contrast

10 Negative or positive image polarity (light characters on a dark background, dark characters on a light background respectively) is acceptable, and each has different advantages. With negative polarity, flicker is less perceptible, legibility is better for those with low acuity vision, and characters may be perceived as larger than they are; with positive polarity, reflections are less perceptible, edges appear sharper and luminance balance is easier to achieve.

11 It is important for the brightness and contrast of the display to be appropriate for ambient lighting conditions; trade-offs between character brightness and sharpness may be needed to achieve an acceptable balance. In

many kinds of equipment this is achieved by providing a control or controls which allow the user to make adjustments.

Screen adjustability

12 Adjustment mechanisms allow the screen to be tilted or swivelled to avoid glare and reflections and enable the worker to maintain a natural and relaxed posture. They may be built into the screen, form part of the workstation furniture or be provided by separate screen support devices; they should be simple and easy to operate. Screen height adjustment devices, although not essential, may be a useful means of adjusting the screen to the correct height for the worker. (The reference in the schedule to adjustable tables does not mean these have to be provided.)

Glare and reflections

13 Screens are generally manufactured without highly reflective surface finishes but in adverse lighting conditions, reflection and glare may be a problem. Advice on lighting is below (paragraphs 20–24).

Keyboard

14 Keyboard design should allow workers to locate and activate keys quickly, accurately and without discomfort. The choice of keyboard will be dictated by the nature of the task and determined in relation to other elements of the work system. Hand support may be incorporated into the keyboard for support while keying or at rest depending on what the worker finds comfortable, may be provided in the form of a space between the keyboard and front edge of the desk, or may be given by a separate hand/wrist support attached to the work surface.

Work desk or work surface

15 Work surface dimensions may need to be larger than for conventional non-screen office work, to take adequate account of:

(a) the range of tasks performed (e.g. screen viewing, keyboard input, use of other input devices, writing on paper);
(b) position and use of hands for each task;
(c) use and storage of working materials and equipment (e.g. documents, telephones).

16 Document holders are useful for work with hard copy, particularly for workers who have difficulty in refocusing. They should position working documents at a height, visual plane and, where appropriate, viewing distance similar to those of the screen; be of low reflectance; be stable; and not reduce the readability of source documents.

Work chair

17 The primary requirement here is that the work chair should allow the user to achieve a comfortable position. Seat height adjustments should accommodate the needs of users for the tasks performed. The schedule requires the seat to be

adjustable in height (i.e. relative to the ground) and the seat back to be adjustable in height (also relative to the ground) and tilt. Provided the chair design meets these requirements and allows the user to achieve a comfortable posture, it is not necessary for the height or tilt of the seat back to be adjustable independently of the seat. Automatic backrest adjustments are acceptable if they provide adequate back support. General health and safety advice and specifications for seating are given in the HSE publication *Seating at Work* (HS(G)57).

18 Footrests may be necessary where individual workers are unable to rest their feet flat on the floor (e.g. where work surfaces cannot be adjusted to the right height in relation to other components of the workstation). Footrests should not be used when they are not necessary as this can result in poor posture.

ENVIRONMENT

Space requirements

19 Prolonged sitting in a static position can be harmful. It is most important that support surfaces for display screen and other equipment and materials used at the workstation should allow adequate clearance for postural changes. This means adequate clearances for thighs, knees, lower legs and feet under the work surface and between furniture components. The height of the work surface should allow a comfortable position for the arms and wrists, if a keyboard is used.

Lighting, reflections and glare

20 Lighting should be appropriate for all the tasks performed at the workstation (e.g. reading from the screen, keyboard work, reading printed text, writing on paper). General lighting – by artificial or natural light, or a combination – should illuminate the entire room to an adequate standard. Any supplementary individual lighting provided to cater for personal needs or a particular task should not adversely affect visual conditions at nearby workstations.

Illuminance

21 High illuminances render screen characters less easy to see but improves the ease of reading documents. Where a high illuminance environment is preferred for this or other reasons, the use of positive polarity screens (dark characters on a light background) has advantages as these can be used comfortably at higher illuminances than can negative polarity screens.

Reflections and glare

22 Problems which can lead to visual fatigue and stress can arise for example from unshielded bright lights or bright areas in the worker's field of view; from an imbalance between brightly and dimly lit parts of the environment; and from

reflections on the screen or other parts of the workstation.

23 Measures to minimize these problems include: shielding, replacing or repositioning sources of light; rearranging or moving work surfaces, documents or all or parts of workstations; modifying the colour or reflectance of walls, ceilings, and furnishings near the workstation; altering the intensity of vertical to horizontal illuminance; or a combination of these. Anti-glare screen filters should be considered as a last resort if other measures fail to solve the problem.

24 General guidance on minimum lighting standards necessary to ensure health and safety of workplaces is available in the HSE guidance note *Lighting at Work* (HS(G)38). This does not cover ways of using lighting to maximize task performance or enhance the appearance of the workplace, although it does contain a bibliography listing relevant publications in this area. Specific and detailed guidance is given in the CIBSE Lighting Guide 3 *Lighting for visual display terminals.*

Noise

25 Noise from equipment such as printers at display screen workstations should be kept to levels which do not impair concentration or prevent normal conversation (unless the noise is designed to attract attention, e.g. to warn of a malfunction). Noise can be reduced by replacement, sound-proofing or repositioning of the equipment; sound insulating partitions between noisy equipment and the rest of the workstation are an alternative.

Heat and humidity

26 Electronic equipment can be a source of dry heat which can modify the thermal environment at the workstation. Ventilation and humidity should be maintained at levels which prevent discomfort and problems of sore eyes.

Radiation

27 The schedule requires radiation with the exception of the visible part of the electromagnetic spectrum (i.e. visible light) to be reduced to negligible levels from the point of view of the protection of users' health and safety. In fact so little radiation is emitted from current designs of display screen equipment that no special action is necessary to meet this requirement (see also Annex B, paragraphs 8–10).

28 Taking cathode ray tube displays as an example, ionizing radiation is emitted only in exceedingly small quantities, so small as to be generally much less than the natural background level to which everyone is exposed. Emissions of ultraviolet, visible and infrared radiation are also very small, and workers will receive much less than the maximum exposures generally recommended by national and international advisory bodies.

29 For radio frequencies, the exposures will also be well below the maximum values generally recommended by national and international advisory bodies for health protection purposes. The levels of electric and magnetic fields are similar to those from common domestic electrical devices. Although much research has been carried out on possible health effects from exposure to electromagnetic radiation, no adverse health effects have been shown to result from the

emissions from display screen equipment.

30 Thus it is not necessary, from the standpoint of limiting risk to human health, for employers or workers to take any action to reduce radiation levels or to attempt to measure emissions; in fact the latter is not recommended as meaningful interpretation of the data is very difficult. There is no need for users to be given protective devices such as anti-radiation screens.

TASK DESIGN AND SOFTWARE

Principles of task design

31 Inappropriate task design can be among the causes of stress at work. Stress jeopardizes employee motivation, effectiveness and efficiency, and in some cases it can lead to significant health problems. The Regulations are only applicable where health and safety rather than productivity is being put at risk; but employers may find it useful to consider both aspects together as task design changes put into effect for productivity reasons may also benefit health, and vice versa.

32 In display screen work, good design of the task can be as important as the correct choice of equipment, furniture and working environment. It is advantageous to:

(a) design jobs in a way that offers users variety, opportunities to exercise discretion, opportunities for learning, and appropriate feedback, in preference to simple repetitive tasks whenever possible (for example, the work of a typist can be made less repetitive and stressful if an element of clerical work is added);

(b) matching staffing levels to volumes of work, so that individual users are not subject to stress through being either overworked or underworked;

(c) allow users to participate in the planning, design and implementation of work tasks whenever possible.

Principles of software ergonomics

33 In most display screen work the software controls both the presentation of information on the screen and the ways in which the work can manipulate the information. Thus software design can be an important element of task design. Software that is badly designed or inappropriate for the task will impede the efficient completion of the work and in some cases may cause sufficient stress to affect the health of a user. Involving a sample of users in the purchase or design of software can help to avoid problems.

34 Detailed ergonomic standards for software are likely to be developed in future as part of the ISO 9241 standard; for the moment, the schedule lists a few general principles which employers should take into account. Requirements of the organization and of display screen workers should be established as the basis for designing, selecting, and modifying software. In many (though not all) applications the main points are:

Suitability for the task

- Software should enable workers to complete the task efficiently, without presenting unnecessary problems or obstacles.

Ease of use and adaptability

- Workers should be able to feel that they can master the system and use it effectively following appropriate training.
- The dialogue between the system and the worker should be appropriate for the worker's ability.
- Where appropriate, software should enable workers to adapt the user interface to suit their ability level and preferences.
- The software should protect workers from the consequences of errors, for example by providing appropriate warnings and information and by enabling 'lost' data to be recovered wherever practicable.

Feedback on system performance

- The system should provide appropriate feedback, which may include error messages; suitable assistance ('help') to workers on request; and messages about changes in the system such as malfunctions or overloading.
- Feedback messages should be presented at the right time and in an appropriate style and format. They should not contain unnecessary information.

Format and pace

- Speed of response to commands and instructions should be appropriate to the task and to workers' abilities.
- Characters, cursor movements and position changes should where possible be shown on the screen as soon as they are input.

Performance monitoring facilities

- Quantitative or qualitative checking facilities built into the software can lead to stress if they have adverse results such as an overemphasis on output speed.
- It is possible to design monitoring systems that avoid these drawbacks and provide information that is helpful to workers as well as managers. However, in all cases workers should be kept informed about the introduction and operation of such systems.

ANNEX B DISPLAY SCREEN EQUIPMENT: POSSIBLE EFFECTS ON HEALTH

THE MAIN HAZARDS

1 The introduction of VDUs and other display screen equipment has been

associated with a range of symptoms related to the visual system and working posture. These often reflect bodily fatigue. They can readily be prevented by applying ergonomic principles to the design, selection and installation of display screen equipment, the design of the workplace, and the organization of the task.

Upper limb pains and discomfort

2 A range of conditions of the arm, hand and shoulder areas linked to work activities are now described as work-related upper limb disorders. These range from temporary fatigue or soreness in the limb to chronic soft-tissue disorders like peritendinitis or carpal tunnel syndrome. Some keyboard operators have suffered occupational cramp.

3 The contribution to the onset of any disorder by individual risk factors (e.g. keying rates) is not clear. It is likely that a combination of factors are concerned. Prolonged static posture of the back, neck and head are known to cause musculoskeletal problems. Awkward positioning of the hands and wrist (e.g. as a result of poor working technique or inappropriate work height) are further likely factors. Outbreaks of soft-tissue disorders among keyboard workers have often been associated with high workloads combined with tight deadlines. This variety of factors contributing to display screen work risk requires a risk reduction strategy which embraces proper equipment, furniture, training, job design and work planning.

Eye and eyesight effects

4 Medical evidence shows that using display screen equipment is not associated with damage to eyes or eyesight; nor does it make existing defects worse. But some workers may experience *temporary* visual fatigue, leading to a range of symptoms such as impaired visual performance, red or sore eyes and headaches, or the adoption of awkward posture which can cause further discomfort in the limb. These may be caused by:

(a) staying in the same position and concentrating for a long time;
(b) poor positioning of the display screen equipment;
(c) poor legibility of the screen or source documents;
(d) poor lighting, including glare and reflections;
(e) a drifting, flickering or jittering image on the screen.

Like other visually demanding tasks, VDU work does not cause eye damage but it may make workers with pre-existing vision defects more aware of them. Such uncorrected defects can make work with a display screen more tiring or stressful than would otherwise be the case.

Fatigue and stress

5 Many symptoms described by display screen workers reflect stresses arising from their task. They may be secondary to upper limb or visual problems but they are more likely to be caused by poor job design or work organization, particularly lack of sufficient control of the work by the user, under-utilization of skills, high-speed repetitive working or social isolation. All these have been

linked with stress in display screen work, although clearly they are not unique to it; but attributing individual symptoms to particular aspects of a job or workplace can be difficult. The risks of display screen workers experiencing physical fatigue and stress can be minimized, however, by following the principles underlying the Display Screen Equipment Regulations 1992 and guidance: that is, by careful design, selection and disposition of display screen equipment; good design of the user's workplace, environment and task; and training, consultation and involvement of the user.

OTHER CONCERNS

Epilepsy

6 Display screen equipment has not been known to induce epileptic seizures. People suffering from the very rare (1 in 10,000 population) photosensitive epilepsy who react adversely to flickering lights and patterns also find they can safely work with display screens. People with epilepsy who are concerned about display screen work can seek further advice from local offices of the Employment Medical Advisory Service.

Facial dermatitis

7 Some VDU users have reported facial skin complaints such as occasional itching or reddened skin on the face and/or neck. These complaints are relatively rare and the limited evidence available suggests they may be associated with environmental factors, such as low relative humidity or static electricity near the VDU.

Electro magnetic radiation

8 Anxiety about radiation emissions from display screen equipment and possible effects on pregnant women has been widespread. However, there is substantial evidence that these concerns are unfounded. The Health and Safety Executive has consulted the National Radiological Protection Board, which has the statutory function of providing information and advice on all radiation matters to Government Departments, and the advice below summarizes scientific understanding.

9 The levels of ionizing and non-ionizing electromagnetic radiation which are likely to be generated by display screen equipment are well below those set out in international recommendations for limiting risk to human health created by such emissions and the National Radiological Protection Board does not consider such levels to pose a significant risk to health. No special protective measures are therefore needed to protect the health of people from this radiation.

Effects on pregnant women

10 There has been considerable public concern about reports of higher levels of miscarriage and birth defects among some groups of visual display unit (VDU) workers in particular due to electromagnetic radiation. Many scientific studies

have been carried out, but taken as a whole their results do not show any link between miscarriages or birth defects and working with VDUs. Research and reviews of the scientific evidence will continue to be undertaken.

11 In the light of the scientific evidence pregnant women do not need to stop work with VDUs. However, to avoid problems caused by stress and anxiety, women who are pregnant or planning children and worried about working with VDUs should be given the opportunity to discuss their concerns with someone adequately informed of current authoritative scientific information and advice.

MANAGEMENT ACTION CHECKLIST 16

Health and Safety (Display Screen Equipment) Regulations 1992 (DSE)

Checkpoints	Action required Yes	No	Action by
Does the company have a copy of the DSE regulations and guidance (indispensable for assessors – see 'Further Reading')?			
Have your staff who use DSE, whether as operators or users, been briefed on these regulations? Did this briefing include the arrangements made for eye tests?			
Did your assessment arrangements include a questionnaire for completion by users and subsequent discussion with them about their perceived problems in working with DSE/VDUs?			
Do your procedures for equipment purchase (DSE and furniture) ensure that nothing is purchased which has not been certified by the manufacturer as complying with these regulations?			
Do all plans for office alterations require a complete review of all the environmental factors covered by these regulations?			
What are your procedures to ensure that there is an assessment review when changes occur to the 'status quo'?			

17

WORKPLACE (HEALTH, SAFETY AND WELFARE) REGULATIONS 1992 (HSW)

As the Workplace (Health, Safety and Welfare) Regulations 1992 (HSW) now contain the 'welfare' requirements for most workplaces in this country, they are described in some depth in this chapter.

INTRODUCTION

The HSW regulations ratify the EU directive on safety and health for the workplace. They came into force on 1 January 1993, but workplaces which existed on 31 December 1992 are allowed a transition period until 1 January 1996 in which to achieve compliance. However, the following workplaces must comply with these regulations from the date they are used:

1. A new workplace opening for the first time after 31 December 1992.
2. Conversion, modification or extension of a workplace which existed before 1993, where the modification or extension commenced on or after 1 January 1993, with the following provisos:
   (a) Where there is a modification or extension, the requirement to comply from the date of bringing the modification or extension into use extends only to that part of the building which has been modified or extended. The unaltered parts do not have to comply until 1 January 1996.
   (b) In the case of a conversion started on or after 1 January 1993, the whole building must comply as soon as it is taken into use (e.g. buildings which are divided into smaller industrial units, a private house or part thereof converted into a workplace, workplaces where significant changes involving structural alterations take place).

Repeals and revocations

Many of the requirements of these regulations repeat the 'welfare' sections of the Factories Act 1961 and the Offices, Shops and Railways Premises Act 1963 respectively; therefore the corresponding sections of these two Acts are repealed, together with a number of orders and regulations which dealt with welfare matters in particular industries, for example the Clay Works (Welfare) Special Regulations 1948.

The regulations therefore 'tidy up' and concentrate workplace welfare requirements, recognizing that whether a person works in an office or industrial premises their welfare needs are similar. Where differences do exist, these are usually significant, and are highlighted in the regulations or guidance.

Status of approved codes of practice (ACOPs)

In common with other chapters covering specific regulations, the information given in this chapter is a combination of the regulations and their associated approved codes of practice (ACOPs) where these exist, and other guidance.

Failure to comply with ACOPs, although not an offence in itself, may be cited where criminal prosecutions are brought for failure to comply with a regulation to which an ACOP refers. It will be for defendants to satisfy the courts that they have met the standards called for in the ACOP in another way.

Exclusions

The following are exempt from compliance with these regulations:

1. *Transport* Ships, boats, hovercraft, aircraft, trains and road vehicles, except that Regulation 13 (see below) applies when an aircraft, train or road vehicle is stationary in a workplace, but not when on a public road.
2. *Mines, quarries and other extractive industries* These subjects have their own legislation to cover the matters addressed by these regulations.
3. *Construction sites including site offices* Where construction is taking place within a workplace, it can be treated as a construction site, provided it is fenced off. Otherwise the site will be subject to these regulations in addition to the Construction Regulations.
4. *Temporary worksites* These include worksites infrequently used and only for short periods, and fairs and other structures which occupy a site for a short period. However, exemption does not extend to Regulations 20–25 which cover sanitary conveniences, drinking water, washing facilities, clothing accommodation, changing facilities and facilities for rest and eating meals, which must be provided 'so far as is reasonably practicable'.
5. *Farming and forestry* Similar to temporary worksites, except that the requirement to make such provision as is 'reasonably practicable' extends only to Regulations 20–22 of these regulations – provision of sanitary conveniences, washing facilities and drinking water.

THE HSW REGULATIONS

REGULATION 4 – REQUIREMENTS OF THE REGULATIONS

The intention is to expand the general duties of employers under Section 2 of the Health and Safety at Work Act (HASAWA) and of persons in control of non-domestic premises under Section 4 of HASAWA, to ensure that *health, safety and welfare* considerations for *all* those who use workplaces are covered. Each of the constituent requirements of the regulations is summarized.

REGULATION 5 – MAINTENANCE OF WORKPLACE, AND OF EQUIPMENT, DEVICES AND SYSTEMS

Regulation 5 requires the workplace and equipment to be properly maintained, kept in efficient working order and in good repair, and properly cleaned. It applies to the following:

1. Equipment, devices and systems which, in the event of a fault or malfunction, would be liable to result in a failure to comply with these regulations.
2. Mechanical ventilation systems provided to comply with Regulation 6, whether or not they include equipment or devices covered by (a) above.

Employers and persons in charge of non-domestic premises are recommended to review their present maintenance systems and arrangements to ensure compliance with this curiously worded requirement.

As a minimum the following items should be the subject of a system of regular maintenance arrangements: emergency lighting, fencing, fixed window cleaning equipment, safety harness anchorage points, devices for limiting window opening, powered doors, escalators and moving walkways.

Systems of maintenance

A suitable system of maintenance should include:

1. Regular maintenance – inspection, testing, adjustment, lubrication and cleaning as appropriate and relevant – carried out at suitable intervals.
2. Isolation and prompt attention to defective equipment.
3. Confirmation that regular maintenance routines are carried out properly and on time, and the maintenance of proper records.

Advice on maintenance and inspection frequencies is as follows:

Equipment The equipment manufacturer's advice should always be sought; other published sources of information emanate from the HSE and British Standards Institution (see 'Further Reading').

Buildings The Chartered Institute of Building Services Engineers (CIBSE) offer advice on building maintenance systems (see 'Further Reading').

REGULATION 6 – VENTILATION

Concerned with the provision of sufficient fresh or purified air to every enclosed workplace, this regulation recognizes that for some industries/processes such provision may not be possible without affecting the process or product. In these cases, workers involved should have adequate breaks in a well-ventilated area.

Mechanical ventilation/AC systems

The importance of regular maintenance and cleaning and the elimination of impurities by filtration is stressed, as is the need to ensure that workers are not positioned where they might be affected by draughts. Where this occurs, some rearrangement of workstations coupled with adjustment of the velocity and direction of air flow should be considered.

If air conditioning system failure could affect health and safety, it should provide a signal to warn of this event. (Although not relevant for most systems, it is important where the system is used to 'dilute' the deleterious effects of dust or fumes or other noxious matter.)

The above applies to general ventilation, and not to local exhaust ventilation for the control of noxious substances (e.g. asbestos, lead), which are the subject of other regulations.

Despite the advance of technology, windows continue to have a key part to play in providing ventilation. But nature's own remedy is often unusable due to external noise and/or the ingress of wind or draught causing paper and so forth to blow around.

REGULATION 7 – TEMPERATURE IN INDOOR WORKPLACES

Ventilation and temperature probably account for more frustration and stress than any other issue at work. One person's draught is another's stuffiness; when the sun shines in winter, half the occupants welcome it, the others want the blinds down.

This regulation does nothing to clarify, nor does it add anything to the sum of, our knowledge. It calls for 'reasonable' temperatures inside buildings during working hours, demands that no heating system shall cause fumes, gas or vapour to enter workplaces where this might be injurious or cause offence, and requires sufficient thermometers to be available to enable staff to check the temperature.

The guidance says that 'reasonable comfort' should be achievable without the need to wear special clothing, but where the processes themselves militate against this, all reasonable steps should be taken to achieve as comfortable a temperature as possible.

This objective is then qualified by the guidance that workroom temperatures should be at least 16°C, or 13°C where the work being done demands severe physical effort. The ACOP concedes that these parameters may not provide 'reasonable comfort' if the air movement or relative humidity is stopping it.

No mention is made of the controls introduced in order to conserve energy which were made under the Fuel and Electricity (Heating) (Control) (Amendment) Order 1980 (S.I. No. 1013). This order limited the use of energy to create a temperature higher than 19°C unless there were medical reasons or there was a need to exceed the maximum in order to ensure that every part of the workplace concerned reached the statutory minimum – 16°C after the first hour of work. There has never been a prosecution for a breach of the maximum, there is no inspectorate to enforce it, and there is almost universal ignorance of the existence of the statutory order! This is fortunate, given that even 19°C is not the most comfortable temperature for those whose jobs make them deskbound for most of the working day!

The HSE are expected to publish more detailed guidance on thermal comfort. Meantime, it is recognized that deficiencies in the heating (and cooling) arrangements of many buildings create a need to use portable heaters or fans. Where even these do not provide optimum thermal conditions, suitable clothing and rest facilities should be considered, perhaps associated with task rotation. Where temporary/portable heating is used, care should be taken to prevent employees being burned, and against the risk of fire.

REGULATION 8 – LIGHTING

This regulation adds little to what was called for in the 1961 and 1963 Acts respectively, that is to say:

1. Every workplace shall have suitable and sufficient lighting, which should, whenever reasonably practicable, be natural light.
2. Suitable and sufficient emergency lighting must be provided in every workroom where its absence would give rise to dangers to those working there in the event that the artificial light failed.

These simply worded requirements mask a whole raft of problems with regard to the highly subjective matter of workplace lighting.

The science of lighting at work has progressed in recent years, and it is now widely recognized that where there is insufficient natural light, there are better solutions than to install glaring overhead lighting. A low background illuminance supported by local (task) lighting is generally preferred, minimizes complaints, reduces eye strain and increases productivity. Uplighters are also a solution to many lighting problems.

Emergency lighting is required for areas which are lit artificially, and where the sudden failure of the mains supply could prove hazardous. Emergency lighting is also necessary for most escape routes. Such lighting should be powered from an independent source, and should come on immediately in the event of mains failure. There is a view that a regular change of all lights at a predetermined frequency offers greater efficiency than changing lights when they fail.

Overhead lighting and skylights should be kept clean. Where skylights create problems of glare or heat gain, this may be overcome by redesigning the workroom layout, or as a last resort by shading the skylights.

Selecting the optimum lighting for a workplace is not a matter for the layman, and obtaining professional advice on lighting matters is invariably cost-effective in the long term.

REGULATION 9 – CLEANLINESS AND WASTE MATERIALS

A permutation of the cleaning requirements of the 1961 and 1963 Acts respectively, this regulation calls for every workplace and its furniture and fittings to be kept 'sufficiently' clean. Floors and walls and ceilings should be capable of being kept sufficiently clean and, so far as is reasonably practicable, waste materials must not be allowed to accumulate, except in suitable receptacles.

The standard of cleanliness to be achieved must be a function of the use to which the workroom is put. For example, medical examination and post-mortem rooms clearly demand much higher standards than factory premises.

There should be a procedure to deal promptly with spillages of all kinds, to attend to leaking pipes and drains, and for the removal of leaked material without delay.

The method of cleaning should not create health and safety risks (e.g. the creation of dusts, the misuse of substances used as cleaning agents).

Where it is not possible to provide suitable receptacles for waste materials, these must be removed from workplaces at least once a day. The novel introduction of waste removal as a specific part of this regulation is to be welcomed. It should signal the end of the practice by some cleaners of assembling waste in plastic bags in fire exits, where it remains throughout the day putting all the building occupants at risk, only to be removed during the evening, when those at risk have all gone home!

However, it would be unfair to suggest that all cleaning problems are of the cleaning contractor's making. One of the most common problems confronting cleaners is that of desks and other workplaces which are littered with paper and other impedimenta. Cleaners are expected to do a thorough job, yet the client's staff leave so much material on their desks that the cleaner often has difficulty in locating the desk top or work surface. Nothing speaks more loudly to a cleaning contractor that the client really does not really care about the appearance of his establishment than masses of materials lying around, and desks, filing cabinets, and so on, which have more paper on top than is housed in the drawers. Apart from the poor standard of cleaning that results, there is also the 'image' question, the danger of industrial espionage – confidential material being left around – and, not least, the increased fire risk created by having combustible material at hand to assist the spread of fire.

REGULATION 10 – ROOM DIMENSIONS AND SPACE

Regulation 10 calls for the provision of sufficient floor area, height and circulation space to allow work procedures to take place without compromising health, safety and welfare.

Some dispensation in this regulation is permitted in that workplaces which existed on 31 December 1992, or are conversions or extensions to such workplaces, will be deemed to meet the requirements of this regulation provided that:

1. They were, until 1992, subject to the provisions of the Factories Act 1961.
2. They do not contravene Part 1 of Schedule 1 of these Workplace Regulations, that is to say:
   (a) no room in the workplace shall be so overcrowded as to cause risk to the health or safety of persons working in it; and
   (b) notwithstanding para. 1, all the persons employed in the workroom shall enjoy a total space of not less than 11 cubic metres, excluding space above 4.2 metres from the floor, and ignoring a gallery if there is one.

Occupants of workrooms should have sufficient space to move about freely. Obstructions – particularly exposed beams in older buildings – must be suitably marked to avoid collision.

The amount of space available in a workroom is affected by the furniture and equipment in it. As a guide, the volume of an empty room when divided by the number normally working in it, should produce at least 11 cubic metres per person, which is a minimum; this will be insufficient if the amount of furniture and equipment in the room is excessive. When calculating, any room more than 3 m high should be counted as 3 m.

REGULATION 11 – WORKSTATION AND SEATING

The ergonomics of work is the subject of this regulation, excluding ergonomics in relation to work with display screen equipment, which is covered by the Health and Safety (Display Screen Equipment) Regulations 1992 (see Chapter 16). The proper ergonomic arrangement of the workstation is required, together with protection from adverse weather when reasonably practicable, the facility to leave the workstation quickly in emergency (or be assisted to do so where appropriate), and elimination of the danger of slipping or falling.

Where work can be done in a seated position, or substantially so, a suitable seat must be provided. Such seating is *not* acceptable unless:

- it is suitable for the person for whom it is provided as well as for the operations performed;
- a suitable footrest is also provided where needed.

As ergonomics are emphasized in this regulation, they should receive special consideration if disabled workers are employed.

Emphasis is placed upon the ability to move about freely, access to materials and controls, minimizing spells where workers are in cramped conditions/positions and, where such work is unavoidable, the provision of a nearby place to recuperate.

REGULATION 12 – FLOORS AND TRAFFIC ROUTES

Ideally all floors and traffic routes should be non-slip, without holes or slopes, and have sufficient drainage where necessary. They must be kept free from obstructions and from anything which might cause persons to slip, trip or fall. Suitable and sufficient handrails, and where appropriate guards, should be fitted to all staircases which are traffic routes, unless they would obstruct free passage on the staircase.

These requirements will be addressed as a matter of course in the safety-conscious and caring company. They not only represent good safety practice, but are the hallmark of an efficient business. Uneven, slippery or obstructed traffic routes are not only dangerous, they create a poor image.

Uneven floors or floors with small holes are often the cause of fork-lift truck accidents, by causing loads to fall off trucks, or worse, the trucks to topple over. Holes and uneven surfaces should be attended to quickly, and suitable warnings, cones or, better, barriers positioned to alert people, particularly in working environments where partially sighted or blind people are present. Prevention is better than cure, and fork-lift truck drivers and other warehouse staff should be alert to deteriorating floors, and report them before they become dangerous. Those carrying out safety audits/reviews should also be especially vigilant with regard to the condition of floors.

Special care is necessary where a slip or fall could be compounded by further injuries (e.g. slipping alongside a machine). Spillage of almost every kind of liquid will make a floor slippery, and a procedure should exist to deal with spillages quickly. Where appropriate, spill containment materials (e.g. sand, granules) should be strategically positioned. Weather can create dangers on floors and traffic routes. Similar procedures to those for spill containment should exist to deal with weather-related hazards, for example snow and ice.

Rapid response arrangements for problems created by indiscriminate parking should be a part of the site security system. Apart from causing obstructions on site traffic routes, such parking can often compromise emergency escape routes or final exits, or obstruct access for fire-fighting or public utility vehicles.

REGULATION 13 – FALLS OR FALLING OBJECTS

The objective of this regulation is to prevent a person falling any distance likely to cause personal injury, or a person being struck by a falling object. The measures taken must, so far as is reasonably practicable, be in addition to the provision of personal protective equipment (PPE) and information, instruction, training and supervision.

Wherever possible, tanks, pits or structures containing dangerous substances, where there is a risk that persons might fall into them, must be securely covered or fenced, as must any traffic route over, across or in them. For the purposes of this regulation, a dangerous substance is: any substance likely to burn or scald; any poisonous or corrosive substance; any fume, gas or vapour likely to overcome a person; or any granular or free-flowing solid substance or any viscous substance which is likely to cause danger.

Fencing

The consequences of falls as described has prompted the requirement for high standards of protection such as secure fencing, which is needed in circumstances where a person could fall two metres or more. Even where falls would be less than two metres, secure fencing is necessary if there are other risk factors: for example, traffic routes pass close to an edge, the danger of falling is increased due to the numbers of people using the route, and so on.

Secure fencing should be of sufficient height and strength. Do not use untensioned chains, ropes or other non-rigid materials.

Covers

Should be strong enough to withstand loads likely to pass over them, especially fork-lift trucks. Covers should be difficult to move or detach.

Temporary removal of fencing/covers Provide secure handholds; make gaps as small as possible; erect suitable signs and barriers where covers are off.

Fixed ladders

Fixed ladders are not a substitute for staircases, but they have a role in certain work tasks (e.g. to facilitate descent into pits, tanks, and so on). If no other handhold exists, ladder stiles should extend a minimum of 1,100 mm above landings – apart from chimneys; ladder stiles should *not* project into the gas stream of chimneys.

Since 31 December 1992, any fixed ladder more than 6 m in height should have a resting place at every 6 m point. Where possible, each run should be out of line with the last, to minimize the distance a person might fall. Where it is not possible to provide resting landings (e.g. on chimneys), ladders should only be used by specially trained, proficient persons.

Where ladders pass through floors, the openings should be as small as possible, and also fenced and gated if reasonably practicable.

Fixed ladders at an angle of less than 15 degrees to the vertical (a pitch of more than 75 degrees) which are more than 2.5 m high should be fitted with safety hoops where possible, or with a fixed-fall arrest system. Hoops should be at 900 mm intervals measured along the stiles, commencing at a height of 2.5 m above the ladder base. The top hoop should be in line with the top of the fencing on the platform served by the ladder. Where a ladder rises less than 2.5 metres, a single hoop should be provided in line with the top of the fencing. Where the top of the ladder passes through a fenced hole in a floor, a hoop need not be provided at that point.

Stairs are safer than ladders, especially when loads are being carried. A sloping ladder is generally easier and safer to use than a vertical ladder.

Roof work

Slips, trips and falls which might be of little consequence at ground level can be fatal if working on a roof. Even when access to roofs is occasional, as with cleaning, inspection or maintenance, it is essential that appropriate precautions are taken.

In addition to falls from roof edges, there is a danger of falling through fragile or worn-out roofing materials. This risk is increased where roofs are subject to bad weather, and where there is pooling due to uneven surfaces or poor drainage.

Where regular roof access is required, suitable permanent access and routes should be defined, with fixed physical safeguards to prevent falling through or off roofs. All fragile roofs should be clearly identified.

Falls into dangerous substances

Every vessel (tank, pit or structure) containing a dangerous substance should be adequately protected to prevent falls into it. Vessels installed after 31 December 1992 should be securely covered, or fenced to a height of at least 1100 mm above the highest point from which people could fall into them. In the case of 'existing' vessels – that is, those *in situ* at 31 December 1992 – the height should be at least 915 mm or, in the case of atmospheric or open kiers, 840 mm.

Changes of level

Changes of level, such as a step between floors which is not obvious, should be highlighted.

Stacking and racking

Racking should be of adequate strength and stability having regard to the loads placed on it and its exposure to damage (e.g. by being hit by fork-lift trucks). Materials should be stored and stacked so that they are not likely to fall. Elements of safe stacking and storage are as follows:

- Safe palletization.
- Banding or wrapping to prevent individual articles falling out.
- Setting limits for the height of stacks to maintain stability.
- Regular inspection of stacks to detect and remedy any unsafe situations.
- Particular instruction and arrangements for irregularly shaped objects.

Loading or unloading vehicles

- Avoid climbing on top of vehicles as far as possible. Where unavoidable, effective measures should be taken to prevent falls.
- Wherever possible, during all vehicle operations (e.g. loading tankers, vehicle sheeting) the work should be carried out from fixed gantries. If this cannot be achieved, those working on top of vehicles should be provided with safety lines and harnesses.

Alternative measures to fencing and covers

- In situations where fences and covers cannot be provided, or have to be removed, special measures must be taken to prevent falls. These include limiting access only to those required to be present for the work, reinforced by barriers. In high-risk situations, a safe system of work, probably including a permit-to-work system, should be implemented. The system might also include provision and use of fall-arrest equipment or safety lines and harnesses and secure anchorage points. Ensure that safety lines are short enough to prevent injury should a fall occur which activates the lines.
- The provision of adequate information, instruction, training and supervision is essential.
- Persons should be prevented from entering any area where, despite the safeguards adopted, there is residual danger, for example from work going on overhead.
- Safety harness/line systems which do not require disconnection should be used in preference to those that do. Where there is no requirement to reach building edges, line lengths and anchorages should be so designed as to prevent the edge being approached.

Scaffolding

Scaffolding and other equipment used for temporary access may either follow the provisions of the approved code of practice (ACOP) or the requirements of the Construction Regulations.

REGULATION 14 – WINDOWS, AND TRANSPARENT OR TRANSLUCENT DOORS, GATES AND WALLS (NEW)

Every window or other transparent or translucent surface in a wall or partition or door or gate must, where necessary for reasons of health or safety, be of safety material to protect it from breakage and be appropriately marked so as to make its presence apparent. The following must be protected:

- Doors and gates, and door and gate side panels, where any part of the transparent or translucent surface is at shoulder level or below.
- Windows, walls and partitions, where any part of the transparent or translucent surface is at waist level or below, except in glasshouses where people should recognize the dangers.

These requirements do not apply to narrow panes up to 250 mm wide measured between glazing beads.

Definition of 'safety materials'

- Materials which are inherently robust, such as polycarbonate or glass blocks.
- Glass which, if it breaks, breaks safely.
- Ordinary annealed glass which meets the thickness criteria in the

following table:

Nominal thickness (mm)	*Maximum size* (m)
8	1.10 × 1.10
10	2.25 × 2.25
12	3.00 × 4.50
15	Any size

An alternative to using safety materials may be to use suitable screens or barriers to prevent persons coming into contact with the glass if they should fall against it. If it is possible for a person to fall down onto glass, the preventative screen or barrier should be difficult to climb.

It is important to mark or identify glass appropriately if this is not made clear by mullions, transoms, rails, door frames, and so on. The marking can be of any design and size provided that it is obvious and at a conspicuous height.

REGULATION 15 – WINDOWS, SKYLIGHTS AND VENTILATORS (NEW)

No window, skylight or ventilator should be capable of opening, closing or adjusting in such a way as to risk the health and safety of the person performing these operations, or when it is open.

The number of accidents in recent years associated with windows, some fatal, highlight the importance of this new regulation. In 1992 a senior executive of American Express was killed after falling from the third-floor window of an office. This accident might have been avoided had the window been fitted with a suitable restraining device.

Methods of reducing risks include the use of window poles, stable platforms, controls placed in such a way as to avoid the danger of falling while using them, or governing the window opening arc to prevent a person falling through the opening. Care should be taken to avoid people colliding with open windows, skylights or ventilators. The bottom edge of opening windows should normally be at least 800 mm above the floor, unless there is a barrier to prevent falls.

REGULATION 16 – ABILITY TO CLEAN WINDOWS SAFELY (NEW)

All windows and skylights must be of a design or construction capable of safe cleaning; equipment used in conjunction with the window or skylight, or devices fitted to the building (e.g. anchorage points), may be taken into account. 'Suitable provision' should be made for windows and skylights to be cleaned if this cannot be done from the ground or another suitable surface. Such 'suitable provision' includes:

1. Windows which are capable of being reversed so that the outside surface is turned inwards.
2. Using access equipment such as suspended cradles, or travelling ladders with a safety harness attachment.
3. Providing 'suitable conditions' for the future use of mobile access

equipment, including ladders up to nine metres long. 'Suitable conditions' include adequate access for the equipment, and a firm level surface in a safe place on which to stand it. Where ladders over six metres long are needed, suitable points for tying or fixing the ladder should be provided.

4. Suitable and suitably placed anchorage points for safety harnesses.

Two factors militate against safe cleaning of windows and glass in many buildings. These are:

1. Finding a safe way to do the job. This is not easy with some modern buildings, where designers and architects have allowed their desire for aesthetic excellence to blind them to the practical difficulties in cleaning the glass.
2. The attitude and approach of window cleaners. Partly as a result of the problems referred to in item 1, preparations to carry out window and glass cleaning take time. Due to the tight schedules to which window cleaners are often held, there is pressure to get the job done quickly; added to this, many cleaners have developed working practices which are inhibited by the use of harness and other equipment.

While the prime responsibility for failure to follow safe systems, and to use the safety equipment provided, must fall upon the window cleaners and their management, the employer/client cannot escape some involvement. However good the cleaning contractor's supervision, it cannot be there all of the time, whereas, unless the windows are being cleaned outside normal working hours, an employer's own staff will be aware of the cleaning operations, and can report unsafe acts or practices.

REGULATION 17 – ORGANIZATION OF TRAFFIC ROUTES (NEW)

1. Every workplace must be organized so that pedestrians and vehicles can circulate safely.
2. The routes must be suitable for the persons or vehicles using them, be sufficient in number and of adequate size.
3. Traffic routes must meet the following criteria:
   (a) pedestrian or vehicle routes can be used as specified without causing danger to persons at work near them;
   (b) there is a separation between vehicle traffic routes from doors or gates or from pedestrian traffic routes which lead onto them; and
   (c) where vehicles and pedestrians use the same traffic route, there is sufficient separation between them.
4. Traffic routes must be suitably indicated.
5. So far as is reasonably practicable, the requirements of item 2 above shall also apply to a workplace which is not a new workplace, modification, extension or conversion.

Traffic routes should be of sufficient width and headroom to allow pedestrians and vehicles to circulate safely and without difficulty. However, for traffic routes which existed before 1993, obstructions such as limited headroom may remain provided there is clear indication of the danger; for example, by marking with hazard tape and if possible by placing padding around the obstruction.

Special consideration is needed where blind or sight-impaired people might use the routes. Persons using wheelchairs also require special consideration. All routes should be sufficiently wide to allow unimpeded wheelchair access, and it may be necessary to provide wheelchair ramps.

Where possible, conventional staircases should exist between floors. Where this is not possible, fixed ladders or steep stairs may be used provided that those who have to use them are capable of doing so safely, carrying loads if this is a requirement of the task.

Care should be taken to ensure that vehicle routes are properly signed, and that all hazards are conspicuously highlighted. Consideration should be given to one-way systems to minimize danger; mirrors at junctions are an important safety feature.

Sensible and realistic speed limits should be imposed, reinforced by limit signs, good supervision, and where necessary, appropriate disciplinary action against offenders. Nothing militates more effectively against safety than speeding which is ignored or even condoned. Speed retarders and road bumps are effective deterrents, but bumps should not be fixed in fork-lift truck routes where this might cause the trucks to shed loads or overturn.

All stanchions, posts, columns and so on in fork-lift truck or other vehicle operational areas should be suitably highlighted and buffered to minimize collision damage.

Workers should be properly protected from exhaust fumes emitting from vehicles.

Vehicle reversing

- Keep it to a minimum or restrict it to a designated reversing area.
- Keep pedestrians and persons in wheelchairs away.
- Provide suitable high-visibility clothing for people who are permitted in the area (e.g. banksmen).
- Fit reversing alarms, and/or detection devices which incorporate automatic brake application.
- Use banksmen.
- Above all, commit the arrangements to a written 'safe system of work'.

Separation of people from vehicles

Wherever possible, vehicle routes should be separated from those for pedestrians. Ground marking of pedestrian routes should be carried out where it is not feasible to effect physical separation.

For buildings in operation before 1993, where there is no vehicle/pedestrian separation and it is not possible to ensure that routes are wide enough for vehicles and pedestrians to pass safely, passing places or traffic management systems should be used.

Crossings

Where there are crossings between pedestrian and vehicle routes, suitable crossing points should be provided and their use enforced. If possible, use guard rails either side of the crossing point to channel pedestrians and prevent risk taking. At points where pedestrians cross vehicle routes there should be an open space of at least one metre between building ends, partitions, and so on and the actual crossing point in order to permit pedestrians to see along the vehicle route in both directions (or in the direction of flow in a one-way system).

Loading bays

These should always have at least one exit point from the lower level, and two for wide loading bays. An alternative is to provide safe 'refuges' to prevent pedestrians being struck or crushed by vehicles. These points should be clearly identified.

Signs

All the routes used by vehicles should carry speed-limit and other signs appropriate to the circumstances. These signs should be the same as they would be on public roads.

REGULATION 18 – DOORS AND GATES (NEW)

Doors and gates must be suitably constructed, including being fitted with any necessary safety devices. They will *not* meet the criteria if:

1. They are sliding doors or gates without a device to prevent them coming off their tracks during use.
2. They are upward-opening doors or gates without devices fitted to prevent them falling back.
3. They are powered doors or gates which are not able to prevent injury by trapping persons.

Furthermore:

- Where necessary for health and safety, any powered door or gate must be capable of opening manually unless it opens automatically in emergency (e.g. if the power fails) – with the exception of lift doors and other doors fitted to prevent falls or access to danger areas. Any tools necessary to manually open a door should be readily at hand. If the power supply is restored during manual operation, the operator should not be at risk.
- Any door or gate capable of being opened by being pushed from either side must provide, when closed, a clear view of the space close to both sides.
- Doors and gates which swing in both directions should have transparent

panels unless it is possible to see over the top. Normal hinged doors on main traffic routes should have viewing panels, capable of being used by persons in wheelchairs.

- Sliding doors should have stops and other fittings necessary to prevent them from coming off their tracks, or from falling down in the event that the suspension system fails.
- Upward-opening doors must be fitted with safety devices (e.g. counterbalance) to prevent them coming down (falling back) on a person.

Power-operated doors and gates must be equipped with features to prevent people being trapped or otherwise injured by them. These include:

1. A sensitive edge, or other suitable detector, and associated trip device to stop, or reverse, the motion of the door or gate when obstructed.
2. A closing-force limiting device, so as to prevent injury.
3. An operating control which must be held in position throughout the door's closing motion. This will only be acceptable where the risk of injury is low and the speed of closure slow. Where such a control is used, it must, when released, cause the door to stop closing or to open immediately; the operator must have a clear view of the door throughout its movement.

Power-operated doors should have readily identifiable and accessible controls to facilitate immediate stopping in emergency. Normal operating controls, if they are positioned properly, will suffice.

Fire resistance

Advice is available from local and fire authorities.

REGULATION 19 – ESCALATORS AND MOVING WALKWAYS (NEW)

These must function safely and be equipped with the necessary safety devices; they must also be fitted with one or more emergency stop controls which are easily identifiable and readily accessible.

REGULATION 20 – SANITARY CONVENIENCES

Suitable and sufficient sanitary conveniences must be provided at readily accessible places. They will not be suitable unless:

1. The rooms containing them are adequately lit and ventilated.
2. They are kept in a clean and orderly condition.
3. Separate rooms are provided for men and women, except where it is possible to lock the conveniences from inside.

If the workplace was in existence on 31 December 1992, or is an extension or conversion, it will be deemed to meet the requirements of this regulation if it was subject to the provisions of the Factories Act 1961, and meets the following criteria:

1. If it is a workplace where females work, there must be at east one suitable WC for use by females only for every 25 females.
2. If it is a workplace where males work, there must be at least one suitable WC for use by males for every 25 males.

When calculating the numbers of males or females working in the workplace, numbers not divisible by 25 count as the next highest number. For example, 35 persons count as 50 persons, and the requirement is therefore for two suitable WCs.

For other workplaces, the requirements are more onerous (see Tables 17.1 and 17.2). The provisions shown in these tables are minimum requirements, and they should be increased if workers take set breaks together, or if they all finish work together and need to wash before leaving work. This is to comply with a requirement of the regulations that facilities must be available 'without undue delay'. Special arrangements are necessary if there are disabled employees who require them.

The basic requirements of WCs are as follows:

- connected to a suitable drainage system;
- provided with effective means for flushing;
- toilet paper in a holder or dispenser;
- coat hook;
- suitable means for disposal of sanitary dressings (women's WC); and
- adequate weather protection.

REGULATION 21 – WASHING FACILITIES

The basic requirements for 'washing stations' are as follows:

- Running hot and cold or warm water.
- Large enough to facilitate effective washing of the face, hands and forearms.
- Showers or baths should also be provided where the work is particularly strenuous, dirty or results in contamination of the skin by harmful or offensive materials (e.g. molten lead in foundries, and in the manufacture of oil cake). Showers which are fed by hot and cold water should be fitted with a device to prevent users being scalded.

Facilities should be arranged to provide adequate privacy, in particular:

- Each WC should be in a separate room or cubicle and be capable of being secured from the inside.
- When any entrance or exit door is opened, it should not be possible to see into urinals or into communal shower or bathing areas.
- Windows of sanitary accommodation, shower or bathrooms should be obscured by means of frosted glass, blinds or curtains unless it is impossible to see into them.

- Facilities should be fitted with doors at entrances and exits unless other measures exist to ensure privacy.

Minimum numbers of facilities

The minimum numbers of facilities that should be provided are shown in Table 17.1, the figures in the first column being the maximum number likely to be in the workplace at any one time. If separate facilities are provided for different groups, either for men and women or for office and manual workers, separate calculations should be made for each group.

Table 17.1 The minimum number of sanitary and washing facilities required in a workplace

Number of people at work	Number of water closets	Number of washstations
1 to 5	1	1
6 to 25	2	2
26 to 50	3	3
51 to 75	4	4
76 to 100	5	5

If sanitary accommodation is used by men only, Table 17.2 may be followed if desired as an alternative to the second column of Table 17.1. A urinal may either be an individual urinal or a section of urinal space which is at least 600 mm long.

Table 17.2 The minimum number of sanitary facilities required in a workplace used only by males

Number of men at work	Number of water closets	Number of urinals
1 to 5	1	1
16 to 30	2	1
31 to 45	2	2
46 to 60	3	2
61 to 75	3	3
76 to 90	4	3
91 to 100	4	4

For numbers above 100, an additional washing station and WC should be provided for every 25 people above 100 (or fractions of 25). Where WCs are used only by men, an additional WC for every 50 men (or fraction of 50) above 100 is sufficient provided at least an equal number of additional urinals are provided.

For work which results in heavy soiling of face, hands and forearms, washing stations should be increased to one for every ten people at work (or fraction of ten) up to 50 people; and one extra for every additional 20 people (or fraction of 20).

Where the provision made for workers is also used by members of the public the number of conveniences and washing stations specified should be increased to ensure that workers can use the facilities without undue delay.

Remote workplaces

For remote workplaces without running water or a nearby sewer, water in containers for washing, or other means of personal hygiene, and sufficient chemical closets should be provided. Chemical closets that require manual emptying should be avoided if possible. If they have to be used, a suitable deodorizing agent should be provided and they should be emptied and recharged at suitable intervals.

Temporary worksites

Suitable and sufficient sanitary conveniences and washing facilities should be provided as far as is reasonably practicable for temporary worksites. WCs and washing stations which satisfy this code should be provided wherever possible. In other cases mobile facilities should be provided which include flushing sanitary conveniences, and running water for washing. The facilities should also meet the other requirements of this code.

Ventilation, cleanliness and lighting

Rooms containing sanitary conveniences should be well ventilated so that offensive odours do not linger. This may be done, for example, by providing a ventilated area between the room containing the conveniences and the other room. Alternatively this may be achieved by mechanical ventilation, or by good natural ventilation, so long as the room containing conveniences is well sealed from the workroom and has an automatic door closer.

No room containing a sanitary convenience should communicate directly with a room in which food is processed, prepared or eaten.

Rooms containing sanitary conveniences or washing facilities should be kept scrupulously clean and well lit. The internal walls and floors should be of a design which permits wet cleaning. Where facilities are shared, responsibility for cleaning should be clearly defined.

REGULATION 22 – DRINKING WATER

An adequate supply of wholesome drinking water must be made available for all employees. It must be readily accessible at suitable places and be clearly marked by an appropriate sign where necessary for reasons of health and safety.

A sufficient supply of suitable cups or other drinking vessels must be provided, unless the supply is from a jet or fountain which persons can use easily. Where the type used are non-disposable, a facility for washing them should be located nearby.

The supply should normally be obtained from a public or private supply by means of a tap on a pipe connected directly to the water main. Alternatively it may be obtained from a tap on a pipe connected directly to a storage cistern which complies with the requirements of UK water by-laws. Cisterns, tanks, and so forth, used as a water supply, should be well covered, kept clean, and tested and disinfected as necessary.

Water should only be provided from refillable containers where it cannot be obtained directly from a mains supply. The containers used must be suitably enclosed to prevent contamination, and should be refilled at least once daily.

Drinking taps should be located where they are free from the danger of contamination, and, as far as is reasonably practicable, not in sanitary accommodation.

Where there is any danger of confusion about which water is suitable for drinking, the drinking water must be clearly marked.

REGULATION 23 – ACCOMMODATION FOR CLOTHING

Suitable and sufficient accommodation for clothing must be provided as follows:

1. For any workers' clothing which is not worn during working hours.
2. For special clothing which is worn by a person at work but which is not taken home.

Such clothing accommodation must meet the following criteria:

1. Where changing facilities are required in accordance with Regulation 24, the accommodation must provide suitable security for the clothing.
2. Where a mix of clothing might create risks to health or damage to the clothing, the accommodation must provide separate accommodation for clothing worn at work and for other clothing.
3. So far as is reasonably practicable, the provision should include facilities for drying clothing.
4. It must be in a suitable location.

Special clothing.

Special clothing is that which is only worn at work (e.g. overalls, uniforms, thermal clothing, and hats worn for food hygiene purposes).

Accommodation for workers' personal and work clothing

The clothing should be able to hang in a clean, warm, dry, well-ventilated place. If the workroom is unsuitable, then accommodation should be in another convenient place. It should consist of, as a minimum, a separate hook or peg for each worker.

Where Regulation 24 applies, effective security measures could be the issue of a lockable locker to each worker concerned.

Where work clothing (including PPE) that is not taken home becomes dirty, damp or contaminated, it should be accommodated separately to the workers' own clothing. Where it becomes wet, there should be a facility to enable it to be dried by the beginning of the next working shift unless fresh dry clothing is provided.

REGULATION 24 – FACILITIES FOR CHANGING CLOTHING

Suitable and sufficient facilities must be provided for any employee who has to wear special clothing for work, and cannot, for reasons of propriety, be expected to change in another room. There must also be separate changing facilities for men and women if this is required, again for reasons of propriety.

Such changing rooms: should be readily accessible from workrooms and eating facilities, if provided; should comprise adequate seating; and should either contain, or communicate directly with, clothing accommodation and showers or baths if these are provided. Privacy should be maintained.

The facilities should be capable of housing the maximum numbers likely to have to use them at any one time without overcrowding or unreasonable delay.

REGULATION 25 – FACILITIES FOR REST AND TO EAT MEALS (NEW)

Suitable and sufficient rest facilities should be provided at readily accessible places. The rest facilities provided must satisfy the following criteria:

1. Where necessary for reasons of health or safety, rest facilities must be provided: in the case of a new workplace, extension or conversion, in one or more rest rooms; or in other cases, in rest rooms or rest areas.
2. They must include suitable and sufficient facilities to eat meals: where food eaten in the workplace would otherwise be likely to become contaminated; and where meals are regularly eaten in the workplace.
3. In rest rooms and rest areas, they must include suitable arrangements to protect non-smokers from discomfort caused by tobacco smoke.
4. They must include suitable rest facilities for pregnant women and nursing mothers.

Each of these requirements are now discussed.

Facilities for eating meals

Suitable and sufficient facilities must be provided for workers to eat meals where meals are regularly eaten in the workplace. Such facilities must also be made available where food would otherwise be likely to become contaminated, including contamination by dust or water. For example:

- cement and clay works, foundries, potteries, tanneries and laundries;
- manufacture of glass bottles and pressed glass articles, sugar, oil cake, jute and tin or terne plates;
- glass bevelling, fruit preserving, gut scraping, tripe dressing, herring curing and cleaning and repairing of sacks.

Smoking

Rest rooms and rest areas must include suitable arrangements to protect non-smokers from discomfort caused by tobacco smoke. This can be achieved either by providing separate facilities or by prohibiting smoking in rest areas.

As the option to prohibit smoking in rest rooms/areas appears now in the approved code of practice (ACOP), this suggests a hardening of attitude. Formerly the advice to employers considering the adoption of a smoking policy was to consult with workers and to give adequate notice of an intention to change the status quo. While it is important to consult with employees in respect to all health and safety matters, the wording of the ACOP gives positive support to employers who wish to ban smoking in rest rooms and rest areas. It follows, therefore, for those workplaces where rest rooms are the only remaining places where smoking is permitted – perhaps due to the dangers of fire, explosion, and so on – that this guidance effectively lays the ground for a total smoking ban.

Facilities for pregnant women and nursing mothers

Suitable rest facilities must be provided for pregnant women and nursing mothers. These should be conveniently sited in relation to sanitary facilities and, where necessary, include the facility to lie down.

It is difficult to envisage a situation where an employer will know in advance – that is, when planning the facility – whether or not a bed or couch will be required. Clearly the wisest course is to provide a bed at the outset.

Rest facilities including seating

Seating Where workers usually stand to do their work, seats must be provided if the work gives them an opportunity to sit from time to time. This requirement existed in the Offices, Shops and Railways Premises Act, but did little for shopworkers, who in many cases – particularly department stores – were enjoined by their employers not to sit down at all as this would create a poor impression!

Seating should also be provided for workers to use during breaks.

If employees normally sit to do their work in a reasonably clean environment (e.g. offices), the normal seating will suffice to meet this duty providing that employees are not subject to excessive disturbance during the rest break (e.g. by intrusive telephone calls, or by enquiries from members of the public).

Personal protective equipment (PPE) Employees required to wear PPE for their work must be able to remove it during rest breaks.

Rest areas or rooms These should be large enough and have sufficient seats with backrests and tables for the numbers likely to use them at any one time. Suitable rest areas must be provided for workers who have to leave their work area and wait until they can return.

The facilities should include the means to prepare or obtain a hot drink (e.g. electric kettle, vending machine, canteen). Where workers are employed during hours or at places where hot food cannot be obtained in, or reasonably near to, their workplaces, they should be provided with the means to heat their own food.

Good standards of cleanliness must be maintained, and care taken that workers do not contaminate the area by substances carried on footwear or clothing. If necessary, adequate washing and changing facilities should be provided nearby.

Canteens or restaurants may be used as rest facilities, provided that there is no obligation to make purchases in order to use them.

CONCLUSION

Despite the scope, size and variety of these regulations, they will, for some businesses, simply confirm that their existing welfare arrangements are adequate; for others, there will be some work to do to achieve compliance.

Generally speaking, a building which was in use before 1 January 1993 has until 1 January 1996 to achieve full compliance with these regulations. The only note of caution to add is that where a non-compliance represents a prima facie breach of existing regulations, or would give rise to serious risk of accident/injury, it is clearly necessary to remedy that defect quickly, and the three-year transition would not apply.

FURTHER READING

Regulation 5

British Standards Institution (1986), *BS 8210: 1986 Guide to Building Maintenance Management,* London: BSI.

Chartered Institute of Building Services Engineers (1987), *Maintenance Management for Building Services*, no. TM17, London: CIBSE.

Chartered Institute of Building Services Engineers (1987), *Operating and Maintenance Manuals for Building Services Installations,* no. BAG1, London: CIBSE.

Health and Safety Executive (1992), *Work Equipment: Guidance on the Provision and Use of Work Equipment Regulations 1992*, no. L22, London: HSE Books.

Health and Safety Executive (1992), *Personal Protective Equipment at Work: Guidance on the Personal Protective Equipment at Work Regulations 1992*, no. L25, London: HSE Books.

Regulation 6

Health and Safety Executive (1988), *Ventilation of the Workplace*, no. EH 22, rev. edn, London: HSE Books.

Health and Safety Executive (1992), *Measurement of Air Change Rates in Factories and Offices*, no. MDHS 73, London: HSE Books.

Regulation 7

Chartered Institute of Building Services Engineers (1986), *Design, CIBSE Guide: Volume A: Design data GVA*, London: CIBSE.

Regulation 8

Health and Safety Executive (1987), *Lighting at Work*, no. HS(G)38, London: HSE Books.
Chartered Institute of Building Services Engineers (1984), *Code for Interior Lighting (CIL)*, London: CIBSE, ISBN 0 900953276.
Chartered Institute of Building Services Engineers (1989), *Lighting Guide: The industrial environment*, no. LG1, London: CIBSE.

Regulation 11

Health and Safety Executive (1991), *Seating at Work*, no. HS(G)57, London: HSE Books.

Regulation 12

Health and Safety Executive (1985), *Watch your step: prevention of slipping, tripping and falling accidents at work*, London: HSE Books.

Regulation 13

British Standards Institution (1982), *BS 6180: 1982 Code of Practice for Protective Barriers in and about Buildings*, London: BSI.
British Standards Institution (1985), *BS 5395: 1985 Code of Practice for the Design of Industrial Type Stairs, Permanent Ladders and Walkways*, London: BSI.
British Standards Institution (1987), *BS 4211: 1987 Specification for Ladders for Permanent Access to Chimneys, other High Structures, Silos and Bins*, London: BSI.
British Standards Institution, *BS 6399: Part 3: 1988 Design Loading for Buildings: Code of practice for imposed roof loads*, London: BSI.
Health and Safety Executive, *Health and Safety in Retail and Wholesale Warehouses*, no. HS(G)76, London: HSE Books. Priced at £7.50.
Health and Safety Executive (1992), *Work Equipment: Guidance on the Provision and Use of Work Equipment Regulations 1992*, no. L22, London: HSE Books. Priced at £5.00.
Health and Safety Executive (1992), *Personal Protective Equipment at Work: Guidance on the Personal Protective Equipment at Work Regulations 1992*, no. L25, London: HSE Books. Priced at £5.00.
British Standards Institution, *BS 1397: 1979 Specification for Industrial Safety Belts, `Harness and Safety Lanyards*, London: BSI.
British Standards Institution, *BS 5845: 1991 Specification for Permanent Anchors for Industrial Safety Belts and Harnesses*, London: BSI.
HMSO, *The Construction (Working Places) Regulations 1966*, (SI no. 94), London: HMSO.
HMSO, *The Construction (Lifting Operations) Regulations 1966*, (SI no. 1581), London: HMSO.
Health and Safety Executive, Guidance Note 6515, *General Access – Scaffolds*, London: HSE Books. Priced at £2.50.
British Standards Institution, *BS 5973: 1993 Code of Practice for Access and*

Working Scaffolds and Special Scaffold Structures in Steel, London: BSI. Priced at £83.

Regulation 14

British Standards Institution, *BS 6206: 1981 Specification for Impact Performance Requirements for Flat Safety Glass and Safety Plastics for Use in Buildings*, London: BSI.

Regulation 16

Health and Safety Executive (1992), *Prevention of Falls to Window Cleaners*, no. GS25, London: HSE Books. Priced at £2.50.
Health and Safety Executive (1983), *Suspended Access Equipment*, no. PM30, London: HSE Books. Priced at £2.50.
British Standards Institution, *BS 8213: Part 1: 1991 Windows, Doors and Rooflights: Code of practice for safety in use and during cleaning of windows (including guidance on cleaning materials and methods)*, London: BSI.

Regulation 17

Health and Safety Executive (1992), *Road Transport in Factories and Similar Workplaces*, no. GS9, rev. edn, London: HSE Books.
Health and Safety Executive (1993), *Safety in Working with Lift Trucks*, no. HS(G)6, rev. end, London: HSE Books.
Health and Safety Executive (1980), *Container Terminals: Safe working practice*, no. HS(G)7, London: HSE Books.
Health and Safety Executive (1985), *Danger! Transport at Work*, no. IND(G)22L, London: HSE Books.
Health and Safety Executive (1988), *Safety in Docks: Docks Regulations 1988: Approved code of practice with regulations and guidance*, no. COP 25, London: HSE Books.

Regulation 19

Health and Safety Executive (1983), *Safety in the Use of Escalators*, no. PM 34, London: HSE Books. Priced at £1.00.
Health and Safety Executive (1984), *Escalators: Periodic thorough examination*, no. PM 45, London: HSE Books. Priced at £2.50.
Health and Safety Executive (1989), *Ergonomic Aspects of Escalators Used in Retail Organizations*, no. CRR12/1989, London: HSE Books. Priced at £40.00.
British Standards Institution, *BS 5656: 1983 Safety Rules for the Construction and Installation of Escalators and Passenger Conveyors*, London: BSI.

Regulation 25

Health and Safety Executive (1992), *Passive Smoking at Work*, no. IND(G)63L, rev. edn, London: HSE Books.
Health and Safety Executive (1989), *Occupational Health Aspects of Pregnancy*, no. MA, London: HSE Books. Free from local HSE offices.

MANAGEMENT ACTION CHECKLIST 17

Workplace (Health, Safety and Welfare) Regulations 1992 (HSW)

Checkpoints	Action required Yes	No	Action by
Has any review action been taken to establish what is necessary to achieve compliance with these regulations by 1 January 1996? (NB: Space will be among the most difficult and emotive areas to get right.)			
In regard to Regulation 25, which includes the duty to segregate smokers from non-smokers in rest areas, has the company taken this into account when considering any proposals relating to smoking policy?			
Where any proposals for new premises, or the upgrade/refit/extension of existing premises, are under consideration, have the requirements of these regulations been taken into account?			

18

MANUAL HANDLING OPERATIONS REGULATIONS 1992 (MHO)

The Manual Handling Operations Regulations 1992 (MHO), made under the Health and Safety at Work Act, are one of the six sets enacted to ratify the EU health and safety directives which took effect from 1 January 1993. They are the first such regulations to deal exclusively with manual handling in all workplaces in the UK. There is no transition period for compliance, as with some other sets of regulations in this group (e.g. DSE, PUWER).

Although manual handling operations rarely cause fatal injuries, they do account for more than 25 per cent of all accidents reported to the enforcing authorities each year. Most of the handling accidents reported are of the 'over-three-day injury' type, and relate to strain or sprain of the back. Frequently these injuries are a cumulative effect of manual handling over a long period rather than a consequence of a particular lifting or carrying task. Complete recovery does not always occur, and there may be physical impairment or permanent disability.

Statistics show that manual handling injuries are not confined to industry, but are widespread. Banking, finance, retail distribution, medical, veterinary and health services all have a high proportion of injuries of this type.

ERGONOMICS

The importance of an ergonomic approach to manual handling is now widely recognized, and the EU directive and these regulations concentrate on ergonomics rather than imposing prescriptive criteria.

For example, they contain no lifting weight maxima, emphasizing instead that manual handling considerations include task, weight, shape and centre of gravity of the load, and, of course, the individual physical capabilities of those involved. However, some industry groups have published guidance that includes recommended maximum weights for manual handling.

MANUAL HANDLING RISK ASSESSMENT

The initial identification of risks associated with materials handling should normally arise from the overall risk assessment which every employer must carry out to comply with the Management of Health and Safety at Work Regulations (MHSW) (see Chapter 15). Handling risks identified during this wide-ranging risk assessment will, where appropriate, identify the need to concentrate on lifting hazards, and a more detailed manual handling assessment must then be carried out in accordance with the MHO Regulations. It is important to cross-reference these activities.

Terms used

Manual handling operations This term means any transporting or supporting of a load (including the lifting, putting down, pushing, pulling, carrying or moving thereof) by hand or bodily force.
Injury For the purpose of these regulations, 'injury' excludes an injury caused by any toxic or corrosive substance which is a constituent part of the load, is present on the surface of the load, or leaks or spills from it.
Load This term relates to the transportation or supporting of virtually any load in a work situation, including humans and livestock.

In fact, the difficulty in developing mechanical solutions to many patient-lifting problems in hospitals, and the unpredictable behaviour of animals, means that minimal benefit will be felt by those engaged in nursing or farming, despite the very high incidence of back-related injury in the nursing profession.

The problems of back injury to nurses and porters in the hospital service are not always a function of financial constraints – although this has an effect in terms of staffing and equipment shortages. Frequently nurses and porters are confronted with situations which demand action at once – and without waiting for assistance or proper lifting aids. At midnight, the solitary nurse on a ward confronted with a patient lying on the floor needs to act quickly. Outside, an elderly person walking into casualty suddenly collapses. Situations of this kind are commonplace. Sophisticated people-lifting aids are expensive, and often in short supply in hospitals. However suitable the aid is, it cannot help the lone nurse with an immediate problem if it is several hundred yards away, in another building, and five floors up.

Other manual handling hazards

Although the emphasis is on preventing back injury, the regulations are concerned with other injuries that might be caused as a consequence of a load's slipperiness, roughness, sharp edges, or due to extremes of temperature.

Combined manual and mechanical handling

The regulations are concerned with the handling of loads by human effort as opposed to mechanical means, although they do apply where, despite the use of mechanical aids, some human effort is involved – for example, where sack trucks are used.

DUTIES OF EMPLOYERS – HIERARCHY OF MEASURES

The duties of employers in this respect are threefold:

1. Avoid hazardous manual handling operations so far as is reasonably practicable.
2. Assess any hazardous manual handling operations that cannot be avoided.
3. Reduce the risk of injury as far as is reasonably practicable. Like all other health and safety considerations, it is for the employer to determine what is 'reasonably practicable'. He may choose not to adopt any further preventative steps when it can be shown that these are disproportionate to the benefits that would derive.

Where employees carry out manual handling away from the employer's premises, the employer should, if possible, liaise with the third parties concerned to ensure that all reasonable measures are taken.

HASAWA also places a duty on employers to look after the health and safety of persons not in their employment while they are on their premises, or where the employer's operations could affect them (see Chapter 1).

AUTOMATION AND MECHANIZATION

If it is established that there is no way to avoid the need for some degree of handling, consideration should first be given to automation or mechanization before resorting to a manual handling solution. Figure 18.1 is a flowchart covering the process.

Mechanization could include the use of lift trucks or conveyor systems, and automating the positioning or movement of a load as part of a process which does not require human intervention. When evaluating these options, account should be taken of any new hazards that they might create (e.g. sudden breakdown, or when undergoing maintenance).

TRAINING

Section 2 of HASAWA imposes a duty on all employers to provide sufficient information, instruction, training and supervision to enable their employees to work safely and without risks to their health. The importance of this requirement

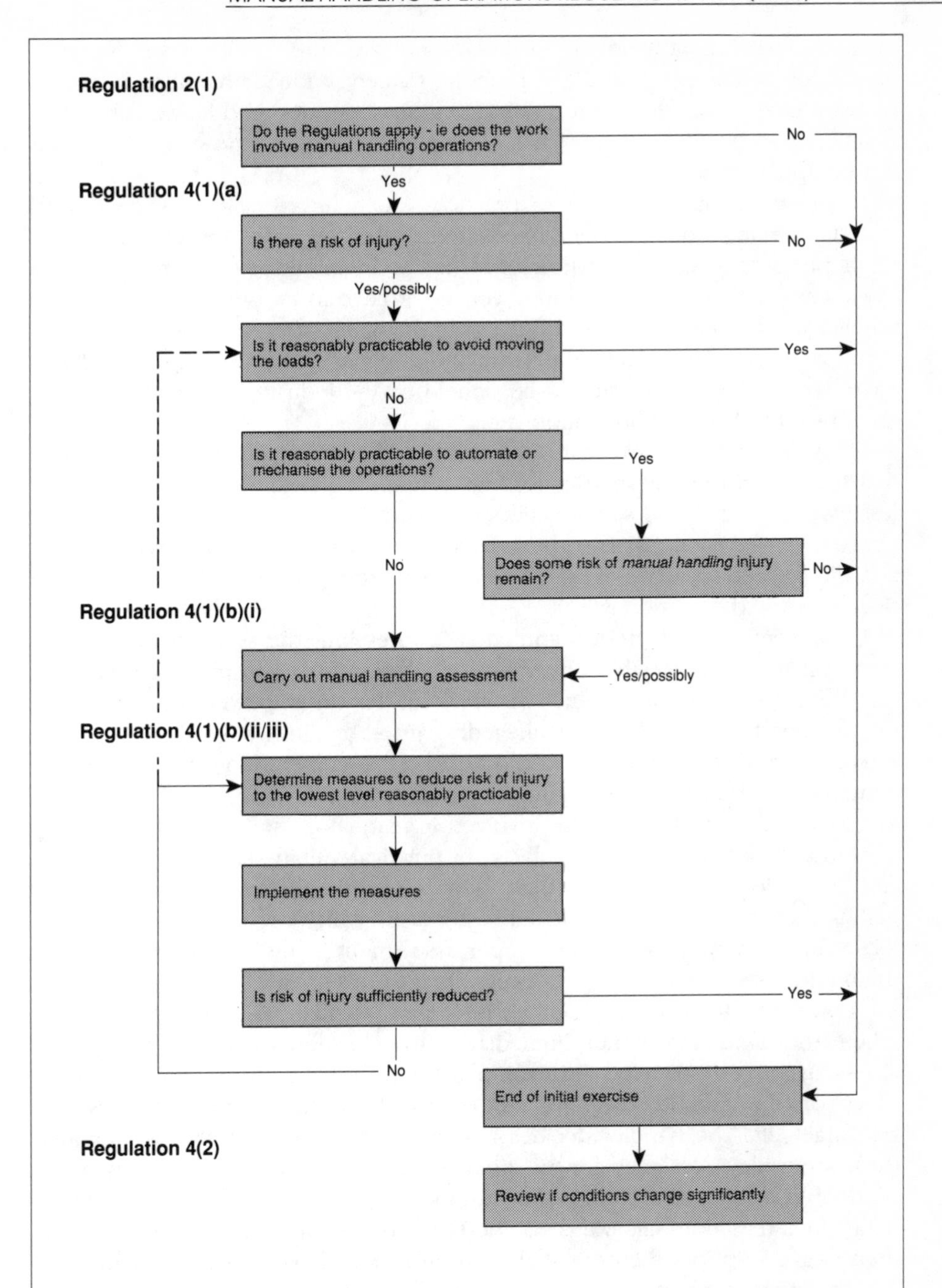

Figure 18.1 Flowchart detailing how to follow the MHO Regulations 1992 (Source: HMSO).
Crown copyright: Reproduced with the permission of the Controller of Her Majesty's Stationery Office.

in relation to manual handling needs no emphasis here. Employees required to handle loads who have not been properly trained in kinetic handling methods, or who have chosen to ignore the training, may develop bad lifting habits and practices which will put them at risk of long-term injury, yet the symptoms might not be apparent at first.

Where continuing poor lifting practices are noted, it will be necessary to repeat the training often enough to ensure that the bad habits are rectified. There might be cases where the poor practices are so ingrained as to be impossible to overcome. In these cases the employer might have to transfer the employee to other work that does not entail lifting, or in the worst cases, consider terminating employment. Provided that the correct disciplinary and warning procedures have been followed, this should be upheld by the industrial tribunal if there is an appeal by the offending employee.

Many civil actions for damages have succeeded where plaintiffs have proved that their employers, despite providing training, did nothing to ensure that the lessons were learnt and safe practices followed. In short, there was no ongoing effective supervision. Worse, there have been cases where supervisors have actually ignored breaches of safe practice, including incorrect lifting, and this has been construed as condonation!

Employers should understand that addressing these regulations in a responsible and thorough manner not only demonstrates compliance with the law, but also serves as a defence in the event that an employee seeks damages or compensation for back injury allegedly caused by lifting at work. Conversely, in such a case, the employer would be in a difficult position if he could not demonstrate a proper approach to the regulations, with evidence of assessment and kinetic training for all those involved in manual lifting.

Manual handling is an ideal subject for practical training in that dummy loads can be made up and awkward situations contrived. Better still, train staff by getting them to carry out the actual lifting tasks which are part of their normal jobs. This training must include proper assessment of the task as described in the model assessment form (see Appendix 2 to this chapter).

Typical of the many problems with regard to manual handling are those where the handler responds immediately to a request for help, without first assessing the job properly. For example, a typewriter is to be moved from an office on the ground floor, to another office on the fourth. There is one passenger lift. The handler decides not to use any mechanical aids, although there are trolleys designed for this kind of task. After passing through two sets of double doors, the handler reaches the lift, only to find that it is being maintained that morning, and is not therefore usable for another hour. He or she then staggers up four flights of stairs, through another pair of double doors, and into the office in which the typewriter is needed, only to discover that every inch of desk space is filled with paper and other machinery! A few moments spent checking out the situation would have avoided a lot of sweat and strain. The old military adage 'Time spent on reconnaissance is seldom wasted' is most apposite here.

ASSESSMENTS (1)

ORGANIZATION AND ARRANGEMENTS

If there is still a need for some manual handling after all options for the elimination of lifting, or the adoption of mechanical or automated solutions, have been considered, the employer must carry out an assessment to quantify the manual handling risks and to take such measures as are reasonably practicable to mitigate them.

Unless the residual manual handling tasks are simple enough to describe verbally to others, the assessment should be in writing or, if stored electronically, be capable of instant retrieval. The HSE have developed a model assessment checklist (see Appendix 2 to this chapter). Assessments should be reviewed whenever handling operations change or other factors demand (e.g. staff changes, in the light of experience).

When considering measures to reduce manual handling, or the risks associated with it, a critical appraisal of existing practices could result in low-cost improvements: for example, a reduction in the size of loads to make them more manageable. This in turn could result in further benefits such as reduced warehouse operation and storage costs. Other possibilities include: rearrangement of storage; replanning traffic routes to offer more space to manoeuvre; specifying team handling; and many others.

Consultation

There is a requirement to consult with workers or their representatives on all matters relating to workplace health and safety. This requirement is most important in respect to manual handling, where the experience of those who have to handle loads is most valuable.

Assessment teams

Manual handling assessments and action plans are matters which ought to be within the capability of a team comprising the following: medical or nursing staff – if employed; the line management of those involved with manual lifting; the company health and safety adviser appointed in compliance with the Management of Health and Safety at Work Regulations 1992 (MHSW) (see Chapter 15); and, of course, representatives of those who carry out manual handling.

The skills/experience mix of the team should include:

- Knowledge of the requirements of these regulations.
- Understanding of the handling operations carried out.
- Basic understanding of human capabilities.
- Identification of high-risk operations.
- Practical ways to reduce risk.

If such experience and skills are not available, or where a company's lifting operations are very large, it might be necessary to bring in outside help, either to supplement the team skills, or to provide training for others (e.g. supervisors) so that they can carry out a risk assessment.

Where employers have opted to appoint an external consultant as safety adviser *vide* the MHSW regulations, and the consultant is not a member of the handling assessment team, he must none the less be involved in a review of the conclusions and action plans.

ASSESSMENTS 2

FACTORS

This chapter is concerned with the practical considerations of manual handling assessments, which are summarized in Appendix 1 to this chapter. Factors which should be considered when assessing manual handling operations are set out in Section B of Appendix 2 under the following generic headings: the tasks; the loads; the working environment; individual capability; and other factors.

When assessing manual lifting tasks, there could be a tendency to disregard those perceived as being marginal in risk terms. This temptation should be resisted. Better to err on the side of caution than to have to defend omissions later.

Kinetic handling training is easy to arrange and inexpensive to conduct. Once the sessions are set up, the opportunity should be taken to put all your employees through the training. After all, the principles apply to *every* lifting task, whether carried out in the workplace or elsewhere.

The trunk – see Figure 2

At what distance from the trunk is the load being held or manipulated? Stress on the back will be in proportion to the distance the load is from the trunk, so keep the load as close to the trunk as possible. Apart from reducing strain and stress, the load is easier to manoeuvre the closer it is to the body (see Figure 18.2).

Posture

Poor posture can result in loss of load control and consequent severe stress on the body. Avoid letting the body weight fall upon the toes with heels off the ground and feet too close together. Never twist the trunk – move the feet in the direction required. Avoid stooping or reaching upwards. Watch for combinations of risk factors (e.g. twisting combined with stooping or stretching, requirement to position loads precisely).

Strains and stresses

Avoid large-reach lifting: that is, do not lift from floor to above waist height in one operation. Avoid excessive carrying distances. When the load is carried further than 10 metres, this, coupled with the stresses involved in the initial lift and final lowering, can be damaging.

Pushing and pulling loads can be injurious if these actions are carried out

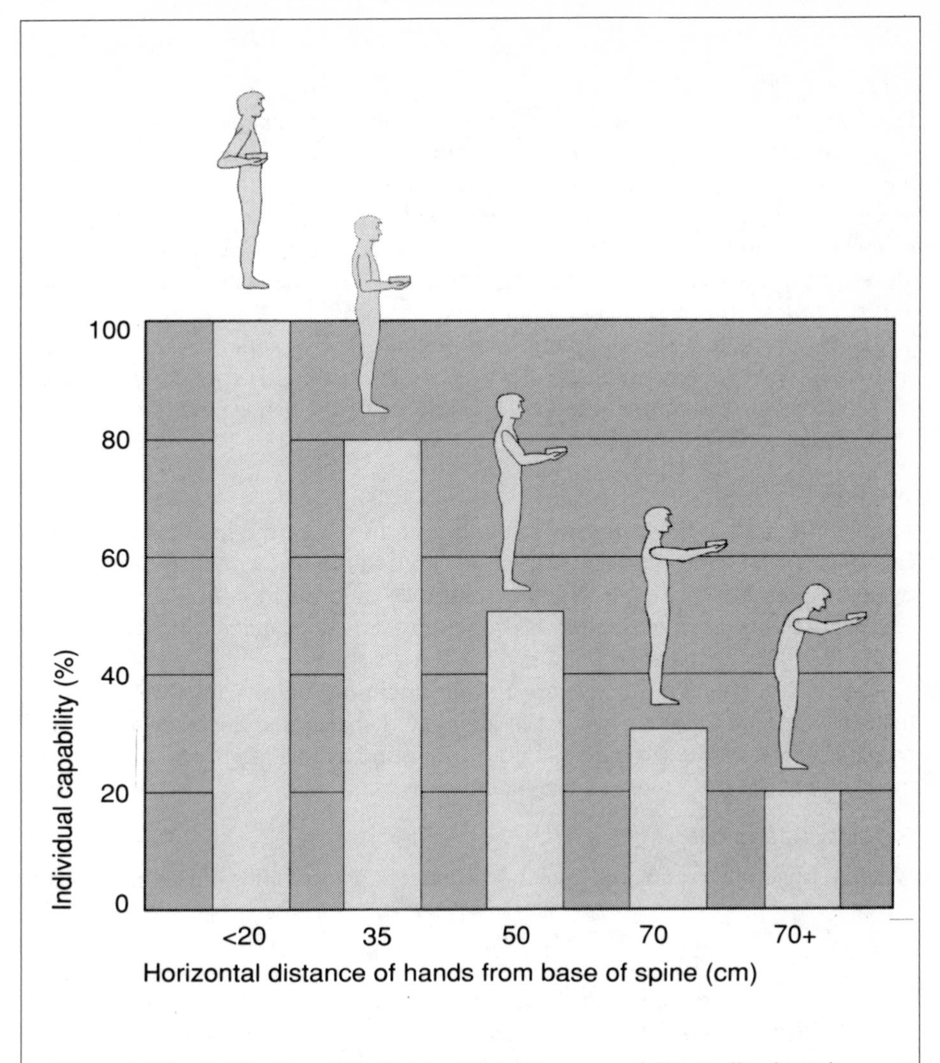

Figure 18.2 Reduction of individual handling capability as the hands move away from the trunk (Source: HMSO).
Crown copyright: Reproduced with the permission of the Controller of Her Majesty's Stationery Office.

with the hands below knuckle height or above shoulder height. They can also be dangerous where foot grip to floor is not certain due to floor or footwear condition.

Is there a risk of sudden load movement which could cause injury due to lack of control of the load (e.g. as in freeing a box trapped on a shelf)?

Is frequent or prolonged physical effort required? If yes, do the arrangements allow for sufficient rest/recovery periods or changes of work activity away from manual handling? Is the handler constrained to work at a pace imposed by the overall work process?

Handling while seated

Handling while seated causes many kinds of additional stresses. The most important aspect is that the powerful leg muscles cannot be used, and therefore the handler's body weight cannot act as a counterbalance. If the load cannot be placed close to the seated handler's body, the stresses of reaching are exaggerated, and there is a danger of the seat parting company with the handler! Moreover, if the seated handler has to lift from below, this will result in stooping and twisting with obvious additional stresses.

These problems are frequently experienced by supermarket checkout operators, where many changes of posture (e.g. reaching below the counter for bags) create strains of many kinds. The advent of the bar-coded checkout has not entirely solved these problems.

Team handling

Although team handling offers a solution to a lift where the load is beyond the capability of a single handler, it is unsafe to assume that two handlers can lift twice the weight of a single handler. A safer equation is to permit a two-person team to carry the sum of two-thirds of their individual capacities, or half the total of the individual capacities if the team is three strong.

Care should be taken to ensure proper communication with team working: for example, that the team are not in a confined space; that they can see around them; and that account is taken of uneven ground as this might result in sudden transfers of weight between team members.

Load considerations

Weight, bulk and centre of gravity, absence of handholds, instability, sharp edges.

Working environment

Poor lighting, space constraints, uneven or slippery or unstable floors, variations in floor level or work surfaces. Thermal problems: for example, difficulty in maintaining grip in very cold conditions, excessive fatigue or sweating in hot conditions, insufficient ventilation, wind making load unstable.

Individual capability

The task requirement might be for handlers with unusual strength, height, and so on. Certainly factors such as sex, age, physique and experience must be taken into account. Occupation too plays an important part: for example, employees working in an environment where lifting is commonplace will be comfortable with lifting tasks that sedentary office workers would find difficult. In general, though, *no* handling task should be attempted if it cannot be satisfactorily performed by most reasonably fit, healthy employees.

Pregnant women/staff with health problems

Where female staff are obviously pregnant or have declared their pregnancy, allowances should be made which should extend throughout the pregnancy and for a period of up to three months after childbirth. Similar consideration should be given to staff known to be in poor health.

CONCLUSION

Although the requirement to make assessments appears onerous, it should be remembered that it is only necessary to undertake them in this degree of detail if handling tasks cannot be eliminated, automated or mechanized, *and* the remaining lifting tasks are such as to constitute a risk when carried out manually. The process can be summarized as follows:

- Is the lifting really necessary anyway – can it be avoided?
- If it is unavoidable, can the frequency be reduced?
- Can the residual lifting work be mechanized or automated?
- If some manual lifting is unavoidable, does it represent a risk to the health of those who will do it?
- If there is a risk, carry out an assessment to decide the safest way to carry out the work.
- Provide sufficient information, instruction, training and supervision as is necessary to avoid injury to those involved.

FURTHER READING

Health and Safety Executive, *Manual Handling – Guidance on Regulations*, no. L23, *Manual Handling Operations Regulations 1992*, London: HSE Books. Priced at £5.00.

Health and Safety Executive, *Lighten the load* information pack, ref. C500, London: HSE Books, containing 'Watch your back' poster, model assessment checklist, *Lighten the Load: Guidance for employers* no. IND(G)109(L), and *Lighten the Load: Guidance for employees* no. IND(G)110(L). Available free of charge from Sir Robert Jones Workshops. Telephone: 051 709 1354/6.

APPENDIX 1

NUMERICAL GUIDELINES FOR ASSESSMENT

INTRODUCTION – THE NEED FOR ASSESSMENT

1 Regulation 3(1) of the Management of Health and Safety at Work Regulations 1992 requires employers to make a suitable and sufficient assessment of the risks to the health and safety of their employees while at work. Where this general assessment indicates the possibility of risks to employees from the manual handling of loads the requirements of the Manual Handling Operations Regulations 1992 (the Regulations) should be considered.

2 Regulation 4(1) of the Regulations sets out a hierarchy of measures for safety during manual handling:

(a) avoid hazardous manual handling operations so far as is reasonably practicable;

(b) make a suitable and sufficient assessment of any hazardous manual handling operations that cannot be avoided; and

(c) reduce the risk of injury from those operations so far as is reasonably practicable.

PURPOSE OF THE GUIDELINES

3 The Manual Handling Operations Regulations, like the European Directive on manual handling, set no specific requirements such as weight limits. Instead, assessment based on a range of relevant factors listed in Schedule 1 to the Regulations is used to determine the risk of injury and point the way to remedial action. However, a full assessment of every manual handling operation could be a major undertaking and might involve wasted effort.

4 The following numerical guidelines therefore provide an initial filter which can help to identify those manual handling operations deserving more detailed examination. The guidelines set out an approximate boundary within which operations are unlikely to create a risk of injury sufficient to warrant more detailed assessment. This should enable assessment work to be concentrated where it is most needed.

5 There is no threshold below which manual handling operations may be regarded as 'safe'. Even operations lying within the boundary mapped out by the guidelines should be avoided or made less demanding wherever it is reasonably practicable to do so.

SOURCE OF THE GUIDELINES

6 These guidelines have been drawn up by HSE's medical and ergonomics experts on the basis of a careful study of the published literature and their own extensive practical experience of assessing risks from manual handling operations.

INDIVIDUAL CAPABILITY

7 There is a wide range of individual physical capability, even among those fit and healthy enough to be at work. For the working population the guideline figures will give reasonable protection to nearly all men and between one-half and two-thirds of women. To provide the same degree of protection to nearly all working women the guideline figures should be reduced by about one-third. 'Nearly all' in this context means about 95 per cent.
8 It is important to understand that the *guideline figures are not limits.* They may be exceeded where a more detailed assessment shows that it is appropriate to do so, having regard always to the employer's duty to avoid or reduce risk of injury where this is reasonably practicable. However, even for a minority of fit, well-trained individuals working under favourable conditions any operations which would exceed the guideline figures by more than a factor of about two should come under very close scrutiny.

GUIDELINES FOR LIFTING AND LOWERING

9 Basic guideline figures for manual handling operations involving lifting and lowering are set out in Figure 18.A1. They assume that the load is readily grasped with both hands and that the operation takes place in reasonable working conditions with the handler in a stable body position.
10 The guideline figures take into consideration the vertical and horizontal position of the hands as they move the load during the handling operation, as well as the height and reach of the individual handler. It will be apparent that the capability to lift or lower is reduced significantly if, for example, the load is held at arm's length or the hands pass above shoulder height.
11 If the hands enter more than one of the box zones during the operation the smallest weight figure should be used. The transition from one box zone to another is not abrupt; an intermediate figure may be chosen where the hands are close to a boundary. Where lifting or lowering with the hands beyond the box zones is unavoidable a more detailed assessment should be made.

TWISTING

12 The basic guideline figures for lifting and lowering should be reduced if the

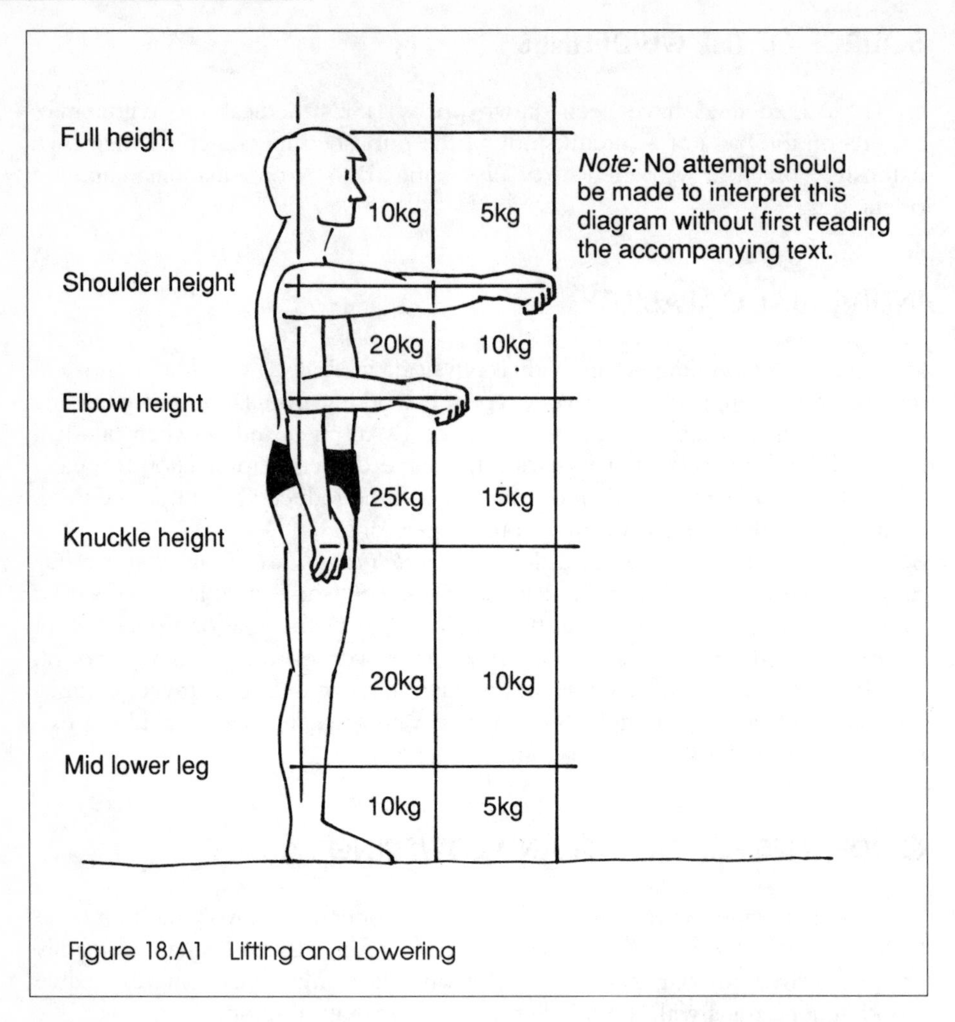

Figure 18.A1 Lifting and Lowering

handler twists to the side during the operation. As a rough guide the figures should be reduced by about 10 per cent where the handler twists through 45° and by about 20 per cent where the handler twists through 90° (see Figure 18.A2).

FREQUENT LIFTING AND LOWERING

13 The basic guideline figures for lifting and lowering are for relatively infrequent operations – up to approximately 30 operations per hour – where the pace of work is not forced, adequate pauses for rest or recovery are possible and the load is not supported for any length of time. They should be reduced if the operation is repeated more frequently. As a rough guide the figures should be reduced by 30 per cent where the operation is repeated once or twice per minute, by 50 per cent where the operation is repeated around five to eight times per minute and by 80 per cent where the operation is repeated more than

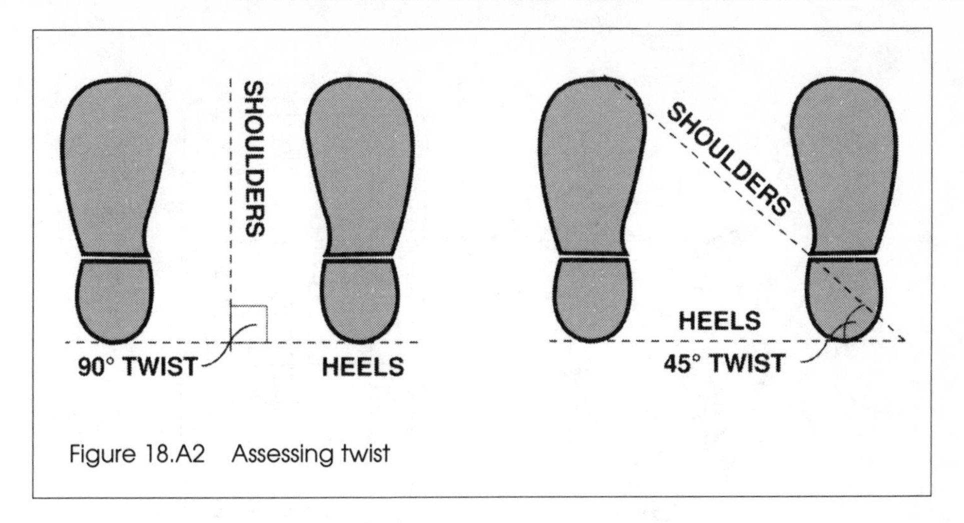

Figure 18.A2 Assessing twist

about 12 times per minute.

GUIDELINES FOR CARRYING

14 Basic guideline figures for manual handling operations involving carrying are similar to those given for lifting and lowering, though carrying will not normally be carried out with the hands below knuckle height.
15 It is also assumed that the load is held against the body and is carried no further than about 10 m without resting. If the load is carried over a longer distance without resting the guideline figures may need to be reduced.
16 Where the load can be carried securely on the shoulder without first having to be lifted (as, for example, when unloading sacks from a lorry) a more detailed assessment may show that it is acceptable to exceed the guideline figure.

GUIDELINES FOR PUSHING AND PULLING

17 The following guideline figures are for manual handling operations involving pushing and pulling, whether the load is slid, rolled or supported on wheels. The guideline figure for starting or stopping the load is a force of about 25 kg (i.e. about 250 Newtons). The guideline figure for keeping the load in motion is a force of about 10 kg (i.e. about 100 Newtons). See Figure 18.A3.
18 It is assumed that the force is applied with the hands between knuckle and shoulder height; if this is not possible the guideline figures may need to be reduced. No specific limit is intended as to the distance over which the load is pushed or pulled provided there are adequate opportunities for rest or recovery.

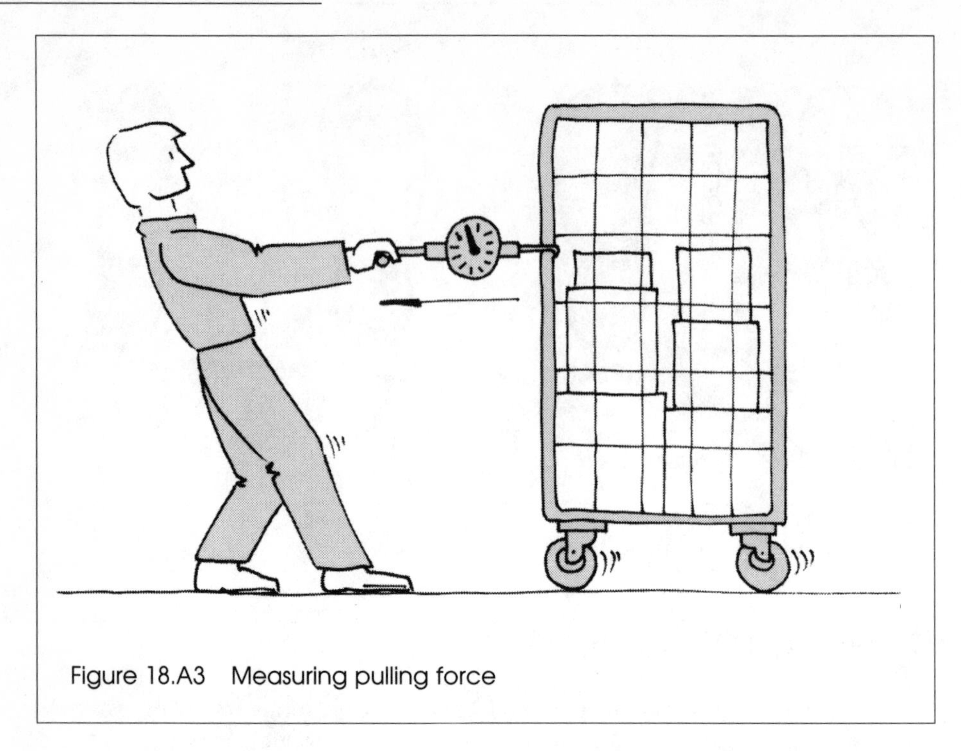

Figure 18.A3 Measuring pulling force

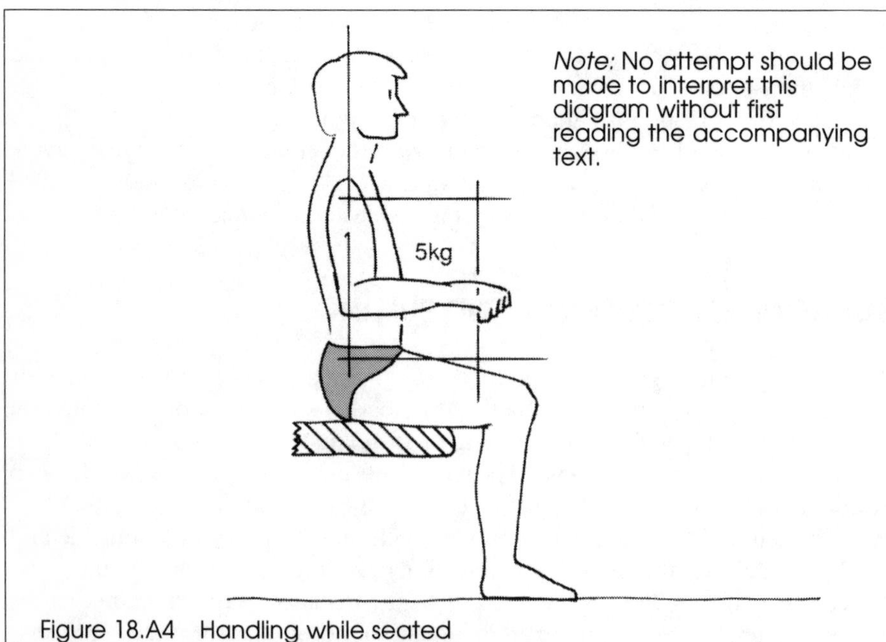

Note: No attempt should be made to interpret this diagram without first reading the accompanying text.

Figure 18.A4 Handling while seated
Remember: the guideline figures should not be regarded as precise recommendations. They should be applied with caution. Where doubt remains, a more detailed assessment should be made.

GUIDELINES FOR HANDLING WHILE SEATED

19 The basic guideline figure for handling operations carried out while seated is given in Figure 18.A4 and applies only when the hands are within the box zone indicated. If handling beyond the box zone is unavoidable or, for example, there is significant twisting to the side, a more detailed assessment should be made.

APPENDIX 2

EXAMPLE OF AN ASSESSMENT CHECKLIST

Manual handling of loads

EXAMPLE OF AN ASSESSMENT CHECKLIST

Note: This checklist may be copied freely. It will remind you of the main points to think about while you:
- consider the risk of injury from manual handling operations
- identify steps that can remove or reduce the risk
- decide your priorities for action.

SUMMARY OF ASSESSMENT	Overall priority for remedial action: Nil / Low / Med / High*
Operations covered by this assessment:	Remedial action to be taken: ....
....	....
....	....
Locations:	Date by which action is to be taken: ....
Personnel involved:	Date for reassessment: ....
Date of assessment:	Assessor's name: .... Signature: ....

*circle as appropriate

Section A - Preliminary:

Q1 Do the operations involve a significant risk of injury? **Yes / No***

If '**Yes**' go to Q2. If '**No**' the assessment need go no further.

If in doubt answer '**Yes**'. You may find the guidelines in Appendix 1 helpful.

Q2 Can the operations be avoided / mechanised / automated at reasonable cost? **Yes / No***

If '**No**' go to Q3. If '**Yes**' proceed and then check that the result is satisfactory.

Q3 Are the operations clearly within the guidelines in Appendix 1? **Yes / No***

If '**No**' go to Section B. If '**Yes**' you may go straight to Section C if you wish.

Section C - Overall assessment of risk:

Q What is your overall assessment of the risk of injury? **Insignificant / Low / Med / High***

If not '**Insignificant**' go to Section D. If '**Insignificant**' the assessment need go no further.

Section D - Remedial action:

Q What remedial steps should be taken, in order of priority?

i ....

ii ....

iii ....

iv ....

v ....

And finally:

- complete the SUMMARY above
- compare it with your other manual handling assessments
- decide your priorities for action
- **TAKE ACTION................AND CHECK THAT IT HAS THE DESIRED EFFECT**

Section B - More detailed assessment, where necessary:

Questions to consider: (If the answer to a question is 'Yes' place a tick against it and then consider the level of risk)	Yes	**Level of risk:** (Tick as appropriate) Low	Med	High	**Possible remedial action:** (Make rough notes in this column in preparation for completing Section D)
The tasks - do they involve:					
◆ holding loads away from trunk?					
◆ twisting?					
◆ stooping?					
◆ reaching upwards?					
◆ large vertical movement?					
◆ long carrying distances?					
◆ strenuous pushing or pulling?					
◆ unpredictable movement of loads?					
◆ repetitive handling?					
◆ insufficient rest or recovery?					
◆ a workrate imposed by a process?					
The loads - are they:					
◆ heavy?					
◆ bulky/unwieldy?					
◆ difficult to grasp?					
◆ unstable/unpredictable?					
◆ intrinsically harmful (eg sharp/hot?)					
The working environment - are there:					
◆ constraints on posture?					
◆ poor floors?					
◆ variations in levels?					
◆ hot/cold/humid conditions?					
◆ strong air movements?					
◆ poor lighting conditions?					
Individual capability - does the job:					
◆ require unusual capability?					
◆ hazard those with a health problem?					
◆ hazard those who are pregnant?					
◆ call for special information/training?					
Other factors - Is movement or posture hindered by clothing or personal protective equipment?					

Deciding the level of risk will inevitably call for judgement. The guidelines in Appendix 1 may provide a useful yardstick.

When you have completed Section B go to Section C.

MANAGEMENT ACTION CHECKLIST 18

Manual Handling Operations Regulations 1992 (MHO)

Checkpoints	Action required Yes	No	Action by
Is a copy of the HSE guidance on MHO held by the company (see appendices to Chapter 18)?			
Is there a nominated individual/steering committee progressing company compliance with MHO?			
Assuming that there are some residual manual lifting tasks after measures to reduce them have been taken, what arrangements exist to explain the dangers, and to provide kinetic training?			
What records are kept? They should include (as a minimum) assessments, either by task or generically, and records of training provided.			

19

PERSONAL PROTECTIVE EQUIPMENT AT WORK REGULATIONS 1992 (PPE)

❖

The Personal Protective Equipment at Work Regulations 1992 (PPE) came into effect on 1 January 1993, being one of the six sets of EU-driven regulations on workplace health and safety. There is no transition period for compliance.

Attitude and culture with regard to personal protective equipment (PPE) has completely turned around since the Industrial Revolution in two important respects.

First, from the standpoint of the employer, the approach was to provide the PPE (sometimes at the worker's expense) and then expect him or her to wear or use it and get on with the job. PPE was the limit of protective measures that many employers would countenance, and in some cases even this was considered to be pampering the workers. Today things are very different, and it is the employer's clear duty *not* to use PPE as the first means of protection, but as a measure to resort to only after a hierarchy of other means have been tried. In fact, in any workplace where PPE is used, the employer will have to demonstrate that all the other options have been considered before turning to PPE as a means of protection.

The second significant change has occurred in the attitude of workers who qualify for PPE. From the outset the trade union movement fought for improved conditions, including those to protect workers' health and safety. Now that the battle has been virtually won, however, the problem is in getting workers to wear the PPE which their forebears fought so hard to obtain!

Much of this reluctance is attributable to: poorly fitting PPE; a perceived failure of the PPE to protect from the hazard for which it was issued; or because the PPE inhibits productivity, which becomes an even more important issue where 'piecework' is the norm. It is also true that in many cases of non-use of PPE, the cause is more a matter of our history and culture. We have a 'cultural' approach to personal safety that makes us uncomfortable about overt displays of concern for our own safety. This is not manifested in a tough attitude, rather it is something which verges upon the apologetic.

Whatever the cause, failure to wear PPE when prescribed has given rise to a considerable volume of case law. This has ranged from successful civil claims where the injured employee has not been issued with PPE when it was clearly necessary, to those where PPE was issued but not worn by the plaintiff, yet the employer did nothing and was therefore judged to have condoned the offence. There have also been successful civil claims where an injured employee plaintiff has been able to prove that he did not wear the PPE provided because it did not fit or did not protect him against the hazard.

There is no doubt that today a criminal or civil action would succeed against an employer if it could be demonstrated that although an injured employee did not wear the prescribed and issued PPE, yet the employer, knowing that it was not being worn, did nothing. In such circumstances the employer *must* take action. This will usually be to invoke the company disciplinary procedure. It would be necessary to pursue this procedure to a final conclusion, which might either be the transfer of the offending employee to other work not requiring PPE, or dismissal. If this course were pursued according to current principles, such an action should be upheld if the dismissal were appealed. Employers simply cannot threaten offending staff where the problem is one of not wearing PPE, and leave the matter unresolved. In one case of this kind the judge concluded that: 'An employer has a duty to remove a stupid employee from the danger which his stupidity places him in.'

It is therefore gratifying to see in Regulations 10 and 11 of these PPE regulations, that employees must not only use PPE issued to them as instructed and look after it, but report forthwith any loss of or obvious defect in any item of that equipment. It would have been better if the wording of Regulation 11 had said, as it could have done, that an employee should report forthwith 'Any loss or obvious defect or *failure to wear PPE for any other reason*'. In subsequent civil actions, however, it should not be possible for any injured employee to allege that the PPE was uncomfortable, did not work, or whatever!

THE REGULATIONS

INTERPRETATION

Personal protective equipment (PPE) means: all equipment (including clothing affording protection against the weather) which is intended to be worn or held by a person at work and which protects him against one or more risks to his health or safety, and any addition or accessory designed to meet that objective.

Exceptions

The regulations do not apply to:

- Ordinary working clothes and uniforms not affording health and safety protection.
- Offensive weapons used for self-defence or as a deterrent.

- Portable detection and signalling devices.
- PPE used for protection while travelling on a road.
- Sports equipment.

There are also exceptions where any of the following regulations apply and in respect to any risk to a person's health and safety for which any of these regulations require the provision or use of PPE, namely:

- the Control of Lead at Work Regulations 1980;
- the Ionising Radiation Regulations 1985;
- the Control of Asbestos at Work Regulations 1987;
- the Control of Substances Hazardous to Health Regulations 1988;
- the Noise at Work Regulations 1989;
- the Construction (Head Protection) Regulations 1989.

WHAT CONSTITUTES PPE?

PPE includes both the following, when they are worn for protection of health and safety:

1. Protective clothing such as aprons, protective clothing for adverse weather conditions, gloves, safety footwear, safety helmets, high visibility waistcoats, and so on.
2. Protective equipment such as eye protectors, life-jackets, respirators, underwater breathing apparatus and safety harnesses.

PROVISION OF PPE

Every employer has a duty to provide PPE where the workplace situation demands, unless the risks have been adequately controlled by other equally or more effective means. A self-employed person has a duty to equip himself with PPE subject to the same qualifications.

PPE will not be suitable unless:

1. It is appropriate for the risk or risks involved and the conditions at the place where exposure to the risk may occur.
2. It takes account of ergonomic requirements and the state of health of the person or persons who may wear it.
3. It is capable of fitting the wearer correctly, if necessary, after adjustments within the range for which it is designed.
4. So far as practicable, it is effective to prevent or adequately control the risk or risks involved without increasing overall risk.
5. It complies with any enactment in Great Britain which implements an EU edict on health and safety in so far as this applies to PPE.

PPE IS A 'LAST RESORT'

The overriding principle in relation to PPE is that it may be used only after a range of other measures have been considered. These measures cover: elimination of the risk; adequate engineering controls; and/or a safe system of work. There may also be a case for issuing PPE even when engineering controls are in place because such controls protect the general environment, not individuals.

CHARGING FOR PPE

Regulation 9 of HASAWA (see Chapter 1) prohibits the levying of a charge for PPE provision.

There have been cases where an employer, although not specifying PPE for a process, has none the less offered employees the opportunity to purchase a form of PPE, usually footwear, at a discount price and through payment by instalments. This prima facie appears unsupportable if it were tested legally. The duty of employers is to determine what PPE is necessary for the task, and then to issue it free of charge.

COMMUNICATION

Communication with employees, important in all health and safety matters, is key in this case. Every employee required to wear PPE should know about these regulations and what his or her employer has done to comply. In particular they should know who the management point of contact is for reporting any problems, and where the PPE is kept, although in most cases PPE will be issued to individuals.

The co-operation and contribution of employees is essential when considering PPE matters. Problems concerning the fit of clothing and equipment, experience when using it, and the extent to which it inhibits work performance, are all matters on which the employee user's contribution is invaluable. Moreover, users are likely to be much more amenable to wearing prescribed PPE if they have been involved in its selection.

Employers should recognize that given the personal nature of PPE – no more racks of communal kit available on a 'first come, first served' basis – there is likely to be a need for a variety of types and sizes of PPE to satisfy a workforce of different shapes and sizes, and with other physical characteristics (e.g. beards).

It is also important to ensure that no reduction in protection occurs when more than one item of PPE has to be worn at the same time, for example a respirator and a hard hat.

ASSESSMENT

An assessment of suitability must be carried out before PPE is purchased, which should include the following considerations:

1. An assessment of any risk or risks to health and safety which have not been avoided by other means.

2. The definition of characteristics which PPE must have in order to be effective against the risks in item 1, taking account of any risks which the equipment itself might create.
3. Comparison of the characteristics of the PPE available with those referred to in item 2.

This assessment must be reviewed if it is no longer valid or there has been a significant change in the matters to which it relates. If as a result of this review, changes in the arrangements for PPE appear necessary, these changes must be made.

The whole of this paragraph on assessments is applicable to both employees and the self-employed.

MAINTENANCE/REPLACEMENT OF PPE

Employers and the self-employed must ensure that the PPE they purchase is maintained in an efficient state, replaced and cleaned as appropriate, and kept in good repair and in efficient working order.

ACCOMMODATION FOR PPE

Where employers or the self-employed issue PPE they must also provide adequate and appropriate storage for it. The accommodation can be simple: for example, a peg for waterproof clothing and helmets; cases for spectacles; or a combination of methods in their vehicle if the PPE is used by workers who are mobile.

INFORMATION, INSTRUCTION AND TRAINING

When an employer issues PPE he must at the same time provide comprehensible training to those required to wear or use it. Such training must cover the following:

1. The risks that the PPE will avoid or limit.
2. The purpose for which the PPE is to be used, and how it is to be worn/operated.
3. The actions necessary by the employee to ensure the PPE is kept clean, in good working order and repair.

FINAL DUTIES

Employers

Having issued PPE, the employer must take all reasonable steps to ensure that it is properly used by employees.

Employee

1. To use the PPE as prescribed and as trained, and to return it after use

to the store provided.

2. To report immediately to his employer or a person specifically nominated by the employer, any loss or defect in the PPE issued to them.

FURTHER READING

Health and Safety Executive, *Personal Protective Equipment at Work – Guidance on the Regulations – Personal Protective Equipment at Work Regulations 1992*, London: HSE Books. Priced at £5.00.

MANAGEMENT ACTION CHECKLIST 19

Personal Protective Equipment at Work Regulations 1992 (PPE)

Checkpoints	Action required Yes	No	Action by
Has your general risk assessment in accordance with Regulation 3 of MHSW (see Chapter 15) identified or confirmed a need for PPE, other than for use to comply with other regulations (e.g. COSHH)?			
If PPE is already prescribed or in use, does it conform to the required standards, and is it maintained and where necessary tested (e.g. RPE)?			
Where employees already need to use PPE, or where your assessment indicates a need to issue it, what arrangements have been made to provide comprehensible information and training?			
Is the provision for accommodating PPE adequate?			
What are the arrangements for reviewing PPE provision in the event of changes in operations, and so on?			

20

PROVISION AND USE OF WORK EQUIPMENT REGULATIONS 1992 (PUWER)

❖

The Provision and Use of Work Equipment Regulations 1992 (PUWER) are in some respects the most curious of all the six sets of EU-driven health and safety regulations enacted here in 1993. First, it is acknowledged that there is virtually nothing new in the regulations, rather that they make more explicit what already exists. Secondly, in so far as equipment being compatible with EU requirements is concerned, there is a fog of confusion and contradiction, which hopefully will clear with the passage of time!

Perhaps the most useful contribution that PUWER makes is that it not only collects up all the old machinery- and equipment-related legislation, but this single regulation also encompasses all places of work.

Provided that employers have been careful to select the best equipment, and have maintained it carefully since purchase, they will have little work to do to comply with these regulations, apart from upgrading some older equipment. Even then there is a transition period until 31 December 1996 in which to do this.

PUWER replaces 17 regulations or parts of them which related to machinery and equipment, although the repealed regulations must remain extant until the last day of the transition period. The expectation is that the general risk assessment which employers must carry out to comply with Regulation 3 of MHSW (see Chapter 15) will highlight any measures necessary to comply with the following parts of these regulations:

- Regulation 5: Suitability of work equipment.
- Regulation 11: Safeguarding dangerous parts of machinery.
- Regulations 12–24: Specific 'hardware' requirements for equipment (e.g. controls, isolation, stability).

To ensure that equipment hazards are in fact included in the general risk assessment, employers' arrangements must emphasize this to those who have any control or responsibility for equipment.

It does not have to be a complicated business; essentially it is a question of deciding whether the equipment, and the arrangements for maintaining it, are adequate to satisfy PUWER, and if not, what additional arrangements are necessary to achieve compliance. In most cases, there will be sufficient 'in-house' expertise to make that decision, and if there are any particularly hazardous or complicated items, specialist advice can be limited to these.

As an aid to this exercise there is a considerable volume of official guidance on the operation and maintenance of equipment, as well as machinery-specific or industry-specific material. If there is no guidance available, or none specific enough, the recognized assessment process should be followed: that is, to decide what is the hazard potential associated with the equipment, how many people could be involved, what is the risk of injury, and how serious this could be?

THE REGULATIONS

TRAINING

The general duty of an employer to provide training appears in Regulation 11 of MHSW (see Chapter 15). Regulation 9 of PUWER is concerned with the content of training as applied to work equipment.

TRADE UNIONS

Safety representatives in unionized companies have a role under these regulations in relation to the selection of equipment and the procedure for maintaining it.

PRECEDENCE OF OVERLAPPING REGULATIONS

Where there is overlap between PUWER and other health and safety regulations, the more specific regulations will apply.

MANUFACTURERS AND SUPPLIERS

The Single Market arrangements for removing barriers to trade have spawned a Machinery Directive which is the 'bible' for manufacturers, and compliance with this directive, ratified here as the Supply of Machinery (Safety) Regulations 1992, will constitute compliance with most of the 'hardware' regulations of PUWER (Regulations 11–24).

RESPONSIBILITIES OF EMPLOYEES

The basic employee responsibilities for health and safety which are contained in Section 7 of HASAWA, have been amplified by Regulation 12 of MHSW (see

Chapter 15). This emphasizes the need for employees to comply with their training, in particular in respect to the use of work equipment and machinery, which must be used correctly.

SUMMARY OF THE APPLICABILITY OF PUWER

The date of application of the regulations varies. The details are contained in Table 20.1.

Table 20.1 Application dates of PUWER

Date of application	Equipment affected
1 January 1993	All new equipment provided for use from this date.
1 January 1997	Work equipment in use before 1 January 1993.
1 January 1993	Second-hand equipment sold by one company to another and brought into use by the second company after 1 January 1993 becomes 'new equipment', even though it is second-hand. It must therefore meet the specific hardware provisions of Regulations 11–24 before being put into use.
1 January 1993	Hired or leased equipment becomes 'new equipment' and must comply with Regulations 11–24 if hired or leased from 1 January 1993 onwards.

'Provided for use'

This means the date on which the equipment was first supplied to the premises. Therefore equipment delivered to a premises before 1 January 1993, but not actually brought into use until after 1 January 1993, is 'existing equipment' and has until 1 January 1997 to reach full compliance.

SCOPE OF THE REGULATIONS

The scope of these regulations is extremely wide. Indeed, it is safer to assume that an item falls within the spectrum, rather than not. Of the hundreds of items covered by PUWER, the following small selection will confirm the breadth of coverage: computers, dumper trucks, ladders, hammers, hand-saws, overhead projectors, cooling towers, pressure vessels. Exclusions include livestock, substances – these are covered by the COSHH regulations (see Chapter 10) – structural items and private cars.

The situation with business transport is, as might be expected, complex. When business vehicles are on public roads, they are subject to the more specific road traffic legislation, which will take precedence. When such vehicles are off public roads, however, these regulations and HASAWA normally take precedence.

The inclusion of hand tools highlights a potential problem of some significance. Where the workers in a particular industry or trade have traditionally used their own (personal) tools for work, this practice might have to end. If the tools are owned by the worker, and used for his employer's business without charge to the employer, the owner could refuse to allow his employer access to them for the purpose of ensuring that they were in all respects safe. This must be an unacceptable position for the employer, who cannot derogate his responsibility under these regulations. Thus a tradition going back to the Middle Ages, and one that has much to commend it, might have to be sacrificed to 'progress'.

THE CONSTRUCTION INDUSTRY

These regulations are difficult for the construction industry to comply with – for example, where more than one contractor might be using the same items of equipment. Some clarification is expected with the forthcoming Construction (Design and Management) Regulations, which will ratify the EU directive on temporary or mobile construction worksites and which is expected to become law sometime in 1994.

SUITABILITY OF WORK EQUIPMENT

This part of PUWER contains the crux of these regulations. It enjoins employers to select the correct equipment for the task, and therefore, by implication, warns of the unacceptable practices of 'making do', or worse, producing a 'lash-up'.

The equipment manufacturers' instructions and guidance should be followed, and factors such as weather, atmospheric conditions, not exceeding the tolerances or capacities of equipment, and the use of equipment in confined spaces, are all matters covered in this section.

MAINTENANCE

Employers must ensure their work equipment is maintained properly, and that records are kept, including completion of a machinery maintenance log if this is provided.

SPECIFIC RISKS

Where a specific risk with work equipment is identified, the employer must:

1. only permit persons specifically nominated to use it;
2. only permit repairs, modifications and maintenance to be carried out by persons specifically designated; and
3. ensure that those nominated in items 1 and 2 have been suitably trained.

INFORMATION AND INSTRUCTION

The need for adequate information and instruction on the following points is reinforced:

1. Conditions in which equipment can be used, and method of use.
2. Foreseeable abnormal situations and action to take.
3. Conclusions drawn from experience with the equipment.
4. Details of these regulations.

TRAINING

The training requirement, including the additional training needed by supervisors, is emphasized.

CONFORMITY WITH EU DIRECTIVES

PUWER places a duty on employers to confirm that new equipment – that is, equipment provided for use for the first time after 31 December 1992 – does comply with EU requirements and bears the 'CE' mark as confirmation of this. This is a 'grey' area, however, and it is not mandatory for *every* item of equipment to carry this mark at present. A purchaser of equipment should therefore take reasonable steps to ensure that the proposed purchase meets the requirements. Such steps could include discussions with the manufacturer/supplier, office of the statutory enforcement body and/or the DTI.

MACHINERY GUARDING

Regulation 11 gathers up the machinery guarding parts of earlier legislation, in particular Regulations 12–16 of the Factories Act.

SPECIFIC 'HARDWARE' TYPE ITEMS

Sections 12–24 cover the specifics, including controls and control systems, isolation from sources of energy, stability, lighting, maintenance operations, and markings and warnings.

FLOWCHART

The flowchart at Appendix 1 to this chapter shows the employer's route to compliance with these regulations in graphic form.

FURTHER READING

Health and Safety Executive, *Work Equipment – Guidance on regulations: Provision and Use of Work Equipment Regulations 1992*, London: HSE Books. Priced at £5.00.

Health and Safety Executive, *Essentials of Health and Safety at Work*, London: HSE Books. Priced at £5.95.

APPENDIX 1

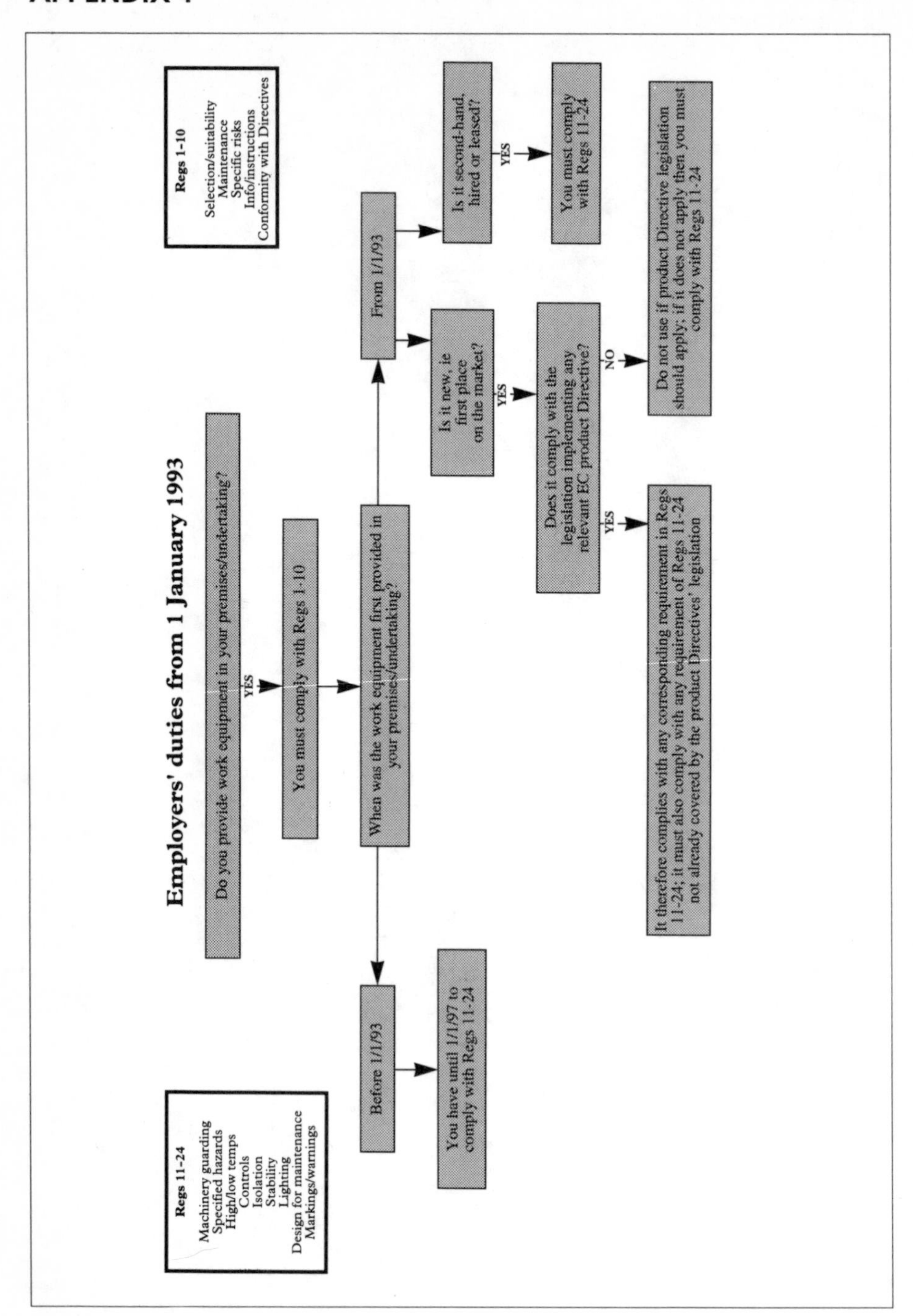
Employers' duties from 1 January 1993
Do you provide work equipment in your premises/undertaking?
YES
You must comply with Regs 1-10
When was the work equipment first provided in your premises/undertaking?
Before 1/1/93
You have until 1/1/97 to comply with Regs 11-24
From 1/1/93
Is it new, ie first place on the market?
YES
Does it comply with the legislation implementing any relevant EC product Directive?
YES
It therefore complies with any corresponding requirement in Regs 11-24; it must also comply with any requirement of Regs 11-24 not already covered by the product Directives' legislation
NO
Do not use if product Directive legislation should apply; if it does not apply then you must comply with Regs 11-24
Is it second-hand, hired or leased?
YES
You must comply with Regs 11-24
Regs 1-10
Selection/suitability
Maintenance
Specific risks
Info/instructions
Conformity with Directives
Regs 11-24
Machinery guarding
Specified hazards
High/low temps
Controls
Isolation
Stability
Lighting
Design for maintenance
Markings/warnings

MANAGEMENT ACTION CHECKLIST 20

Provision and Use of Work Equipment Regulations 1992 (PUWER)

Checkpoints	Action required Yes	No	Action by
Has any formal work been initiated to assess the measures necessary to comply with the work equipment regulations?			
If yes, what is the extent of equipment requiring replacement and at what cost?			
Do managers responsible for equipment know that if they have any equipment that presents a hazard, this should be highlighted in the general risk assessment in accordance with MHSW?			
If any staff use or bring their own tools to work, how will you control this, given that you (the employer) have total responsibility for their condition?			

21

CONSTRUCTION (DESIGN AND MANAGEMENT) REGULATIONS 1994 (CD&M)

This book has been written for the benefit of managers across the spectrum of business; every chapter deals with legislation affecting all workplaces or work situations.

Although this is not strictly the case with the CD&M Regulations, they are included because they have the potential to affect every business at some time or other, and, as the various duties imposed by the regulations show, there are duties which could be assigned to professionals and managers who are *not* normally involved in work of a construction nature.

These regulations came into law on 31 March 1995, somewhat later than required by the European Directive which they ratified – the Temporary and Mobile Worksites Directive, more generally known as the Construction Safety Directive.

The delay in introducing the regulations was in part due to the huge volume of comment during the consultative phase, not only from the professions and disciplines directly involved, but from other interested parties. It is reported that these regulations have generated more comment than for any health and safety related regulations; this has also been the experience of other EU member states.

This might suggest that the regulations are unnecessary or unduly burdensome. While such criticism might be justified in respect of some EU-generated health and safety law, it would be wrong to suggest that the CD&M regulations fall into this category.

The construction industry has been almost synonymous with danger; indeed, when the industry mounted a construction safety campaign some years ago, the year in question was marked by an even greater number of injuries than usual!

Another telling example of the problem of bringing about safety improvements in the industry was the inordinate delay in getting The Construction (Head Protection) Regulations onto the statute book.

For over 200 years an annual toll of death and serious injury due to being hit by objects falling from above did little to focus the minds of those who should have been addressing the problem in a positive manner.

From another perspective, thousands of commercial and industrial buildings and sites are without any record of the construction details: how they should be maintained; what 'as fitted' services there are; what the load-bearing properties of floors are; and much else besides.

Within the industry itself, there are examples of poor co-operation and liaison, especially on major projects, where there may be a number of contractors. There is ample evidence to confirm that in some cases each of the contractors has worked in a 'blinkered' fashion, without regard for what others on the site are doing, or to the need to communicate and co-ordinate.

Some designers, while delighted to receive acclaim for the aesthetic quality of their projects, are not so keen to discuss the problems of maintenance and cleaning – especially the windows of the glass palaces which are now a feature of so many inner city and 'out-of-town' developments. In some cases, even trained monkeys could not manage the contortions needed to clean the glass!

It is not surprising, therefore, that there has been so much input to the consultation process, or that in some cases there have been delaying tactics in the hope that the regulations could be substantially modified, or even abandoned altogether.

Now that the regulations are in place, there is considerable evidence that some of those who have specific duties under them are endeavouring to get the client to pay! Put bluntly, this amounts to a request for extra money to do what they should have been doing as good practice from the outset. Rather like saying 'Oh, you want us to design it/build it safely – well, that will cost a lot more'.

Of course this is an over-simplification. Some additional costs will be incurred, and some of these should be passed on to clients. The reader will be able to judge the extent by studying the specific responsibilities of the various 'dutyholders' under these regulations. It is quite clear that many of these duties are simply matters of good practice – things that any efficient and conscientious firm would expect to do as a matter of professional pride in the job.

There is no doubt that CD&M will not be a palliative for all the problems described in this chapter, and it may take some years before the full benefits of the regulations appear. However, it is a start, and some drastic measures are needed to 'stop the rot'.

The remainder of this chapter is devoted to explaining the roles and responsibilities of what the regulations describe as 'dutyholders'. As stated in the preamble to this chapter, it is important that *all managers* understand these regulations.

At some time in their career, every manager could be involved in the upgrade, refurbishment or renovation of a building, or indeed in the planning of an entirely new facility.

If the scope of this work falls within the ambit of these regulations, they may either become dutyholders – in which event they have specific statutory responsibilities or they might need to understand the responsibilities of those who are directly involved in the project – architects, designers, planning supervisors, principal contractors and other contractors.

DUTYHOLDERS

The regulations stipulate a number of key appointments in relation to a construction project. Each of these appointments is filled by a dutyholder, whose specific responsibilities are clearly defined. These are described below.

CLIENTS (Including clients' agents and developers)

Note: An employer/client will normally appoint an agent to be responsible for the discharge of the duties that follow, and any 'competent' manager/ professional in the organization could become the clients' agent.

DUTIES

- Must be reasonably satisfied that they appoint competent persons as Planning Supervisor, Designer and Principal Contractor.
- Must be satisfied that sufficient resources (time/financial) have been, or will be, allocated to carry out the project safely.
- Must provide the Planning Supervisor with all necessary information relating to the site on which the project will take place.
- Once the project is completed, the client must receive and take custody of the Safety File from the Planning Supervisor. It is essential that this file is maintained and updated when appropriate, so that it remains a 'living document' which is, effectively, the 'building bible' (see paragraph titled 'Health and Safety File').

Notes:

i Domestic construction work does not fall within the ambit of the CD&M Regulations.

ii Responsibility for ensuring the competence of the dutyholders whom he appoints rests with the client, who must, therefore, make such checks and enquiries as he deems necessary to discharge this duty properly.

Competence is not qualified in absolute terms. Generally speaking, a person will be competent if they have the necessary experience or training in the subject, and have the right temperament/attitude which in this case means a positive and committed approach to health and safety in addition to their professional competence.

Comment: For many clients, the above requirements will come as something of a shock. Not only must they ensure that they appoint competent dutyholders, but also that sufficient resources – of time and money – are available to carry out the project safely and without risks to health. The days when a small sum held as 'contingency' in case it was needed for a safety-related problem are over!

PLANNING SUPERVISOR

An important new position, whose incumbent has overall responsibility for co-ordinating the health and safety aspects of the design and planning phases of the project.

DUTIES

- To ensure that a Health and Safety Plan is prepared (see paragraph titled 'Health and Safety Plan').
- To monitor the health and safety aspects of design.
- To ensure co-operation between all designers.
- To advise the Client on the allocation of health and safety resources, and the competency required of the designer.
- To prepare the Health and Safety file.
- To advise the Client in respect of the Principal Contractor (as to competence, Health and Safety resources and attitude, etc).
- To notify the project to the HSE if the work is likely to exceed 30 days or 500-person days of construction work.

DESIGNER

Generally, the duty of designers is to design in a way which avoids, reduces or controls risks to health and safety as far as is reasonably practicable to enable projects which they design to be constructed *and maintained* safely.

Unresolved risks must be identified to enable the construction contractor to operate safely.

PRINCIPAL CONTRACTOR

The Principal Contractor is in charge of the construction site.

DUTIES

- To take over, develop, implement and monitor compliance with the Safety Plan.
- To co-ordinate health and safety activity on site.
- To make the site secure.
- To arrange for competent and properly resourced sub-contractors.
- To obtain details of sub-contractors' risk assessments.
- To provide information to everyone on site.

- To ensure that adequate training has been given to everyone on site.
- To enforce compliance with site safety rules.
- To ensure that all workers are properly informed and consulted on health and safety matters.
- To display the notification of the project to the HSE.
- To pass information to the Planning Supervisor for incorporation into the Health and Safety file.
- To ensure that all necessary safety signs, notices, etc are properly displayed.

CONTRACTORS (Employers and the self-employed)

DUTIES

- To co-operate with the Principal Contractor, including advising him of all risks associated with their work, and informing him of any RIDDOR reportable accidents, etc.

GENERAL NOTE ON DUTYHOLDERS

The CD&M Regulations do not require that the dutyholder posts described are filled by different persons. If the client is satisfied as to the competence of those to be appointed, he may opt to assign more than one of the dutyholder's posts to an individual.

THE HEALTH AND SAFETY PLANS, HEALTH AND SAFETY FILE AND OFFICIAL NOTIFICATION OF A PROJECT

NOTIFICATION OF A PROJECT

Each of the above three items are statutory requirements if the project falls within the ambit of these regulations, ie it lasts for more than 30 days or will involve more than 500-person days of work.

If a project does not exceed this threshold it does not have to be notified, although it must comply with the CD&M Regulations if the total number working on site at any time reaches five or more. All design work is subject to CD&M Regulations irrespective of the size of the project, as is a project which involves demolition.

THE HEALTH AND SAFETY PLAN

The Health and Safety Plan is intended to ensure that the construction phase of the project is carried out safely. The plan should be initiated by the Planning Supervisor *before* tenders are sought, so that the tendering contractors know what is required.

Once the Principal Contractor is appointed, it is his duty to develop the plan to reflect the changing circumstances as the construction phase proceeds, keep it up to date, and ensure compliance with it by all those on site.

THE HEALTH AND SAFETY FILE

The Health and Safety File is intended to be the 'corporate memory' of the building, which will be altered/amended and updated on a continuing basis.

Initially the file should be prepared by the Planning Supervisor and incorporate any information provided to him by the Client. Subsequently amendments will be made as a result of input from the Principal Contractor. When the project is completed, the file will be passed to the Client, whose duty it is to maintain/update the file throughout his ownership of the building.

It must pass to any new owner, and be kept available until the building is finally demolished. The format of the file will vary according to the size of the building, and it might be the case that the file simply records where the necessary information can be located.

Effectively the file is the history of the building from inception to demolition, and will contain at least 'as-built'/'as-fitted' drawings, maintenance manuals, records of servicing, details of any re-arrangements, refurbishing, etc.

FURTHER READING

HSE Approved Code of Practice (ACOP): Managing Construction for Health and Safety. Construction (Design and Management) Regulations 1994. ISBN 0–7176–0792–5 Price £7.95. HSE BOOKS Reference L54

HSE PAMPHLET CDM REGULATIONS — How the regulations affect you! Reference PML54. Free

MANAGEMENT ACTION CHECKLIST 21

Construction (Design and Management) Regulations 1994 (CD&M)

Checkpoints	Action required Yes	No	Action by
What process exists in the company for ensuring compliance with the CD&M Regulations when proposed projects fall within the regulations?			
What training in the CD&M Regulations needs to be provided for *a)* staff in planning/property services/real estate or similar departments, and *b)* all other managers?			
Key dutyholders must be formally appointed in writing, even if they are employees. What arrangements have been made to do this for projects falling within CD&M Regulations, and has the letter of appointment draft been cleared by the legal department/advisors?			

Reproduced from *What Every Manager Needs to Know About Health and Safety* by Ron Akass, Gower, Aldershot, 1995.

22

THE WAY FORWARD

The considerable publicity which surrounded the emergence of the European Single Market had a number of 'spin-off' effects, one of which was to increase awareness of the role of occupational health and safety in the order of things. Nevertheless 'awareness' does not necessarily mean support, and the danger that employers will become cynical at the non-stop flow of EU-driven legislation of all kinds is very real indeed.

As the bombardment of directives continues, the standard words of the preamble to every EU directive have an increasingly hollow ring: 'the requirements must avoid imposing administrative, financial and legal constraints in a way which would hold back the creation and development of small and medium-sized undertakings'. However sincere the intention, the risk is that the health and safety message and ethos could be weakened because of the volume of legislation.

At the other extreme, there are those who have received a very different kind of signal from all the health and safety legislation. They believe that standards must have been very low to require six new sets of health and safety regulations to put them right!

Of course, these regulations were made to ratify EU directives, which in turn were enacted to address shortcomings in health and safety standards in the Union as a whole.

January 1993 could not have been a less propitious time to introduce *any* new laws. Europe was in recession, and the problem for many businesses was survival, not spending money and resources responding to new directives. In the midst of this uncertainty for most companies, two areas of business were booming. One was the health and safety consultancy, the other those manufacturers whose products were or would be in demand as businesses started to move toward compliance with the new regulations.

The Health and Safety Executive felt it necessary to publish guidance to employers titled *Selecting a Health and Safety Consultancy*, realizing, no doubt, that however hard they tried to minimize the resource and cost implications of the new and revamped requirements, the regulations themselves, together with their associated guidance, gave the lie to the notion that there really was not a lot to do to achieve compliance. What the legislators and enforcers appear to have overlooked is the fact that there are no longer odd employees waiting around for something to do!

THE WAY AHEAD

There is, however, a real chance that good will come out of these circumstances. Existing legislation is under scrutiny, and it is to be hoped that the lessons of '1993' will be learned by the Brussels bureaucracy, with the result that there will be a slowing down and rationalization of directives in the future.

In the meantime, for many companies, increasing awareness of the importance of having a *safe and healthy* workforce has led to a variety of initiatives which would have been undreamed of even thirty years ago. In addition to regular medical examinations, there are health and fitness classes – even in a few cases gymnasiums and swimming pools – health advisory services, including counselling on drug and alcohol problems, and other commendable initiatives.

The smoking issue, and the concern of employers to address this effectively, has given rise to a variety of solutions, some of which have been challenged by disgruntled employees at industrial tribunals. Generally the outcome has been in favour of the move toward smoking reduction or cessation, although in some cases the timescale or procedures followed by an employer have been criticized. The dilemma for many employers is in trying to take account of the wishes of smokers and non-smokers while government remains 'on the fence', unable or unwilling to legislate – which they will eventually be forced to do. It will be surprising, though, if the EU produces anything positive on this subject, given the high incidence of smoking in most member countries.

MANAGERS AND EXECUTIVES

The impetus, then, for managers and executives to become more closely involved in the business of health and safety is unstoppable. Where their involvement and commitment is not forthcoming as a matter of commonsense and good business, there is statutory direction to enforce it, and the substantial increase in sanctions available to magistrates, including that of imprisonment for some offences, gives increased effect to health and safety law.

Yet although examples of wilful disregard for the health and safety of employees occur less often, there are still too many serious accidents in which poor management control or the absence of safe systems of work were the underlying cause of the tragedy.

Those who use the checklists in this book to establish whether their subordinates or the company are doing all that they should to further the cause of health and safety will make an even more important discovery in the process – namely, whether they are themselves setting a good example, and giving the lead to those with whom they interact at work.

Each of us has a difficult transition to make. History has left us with a legacy of reticence when it comes to overt displays of concern for our own safety. We have to find a way to overcome this at work; to accept that taking care is not in any sense feeble or ineffective. Perhaps those of us who recognize these traits in ourselves can work out our energies to the limit in the gym or on the sports field. In this way we can compensate for taking more care and being more concerned with health and safety at work.

And thus the objectives of *health and safety* will be achieved both in and out of work!

INDEX